AQA Design and Technology

Systems and Control Technology

GCSE

David Larby

Chris Sherwood

Mark Walton

Nelson Thornes

Published in 2009 by:
Nelson Thornes Ltd
Delta Place
27 Bath Road
CHELTENHAM
GL53 7TH
United Kingdom

13 14 15 16 / 10 9 8 7 6 5 4 3

A catalogue record for this book is available from the British Library

ISBN 978 1 4085 0418 5

Cover photograph: Jim Wileman
Page make-up by Fakenham Photosetting Ltd
Additional artworks by Fakenham Photosetting Ltd and Angela Knowles
Printed and bound in Spain by GraphyCems

Contents

Nelson Thornes has worked hard to ensure this book and the accompanying online resources offer you excellent support for your GCSE course.

You can feel assured that they match the specification for this subject and provide you with useful support throughout your course.

These print and online resources together **unlock blended learning**; this means that the links between the activities in the book and the activities online blend together to maximise your understanding of a topic and help you achieve your potential.

These online resources are available on which can be accessed via the internet at **www.kerboodle.com/live**, anytime, anywhere. If your school or college subscribes to **kerboodle** you will be provided with your own personal login details. Once logged in, access your course and locate the required activity.

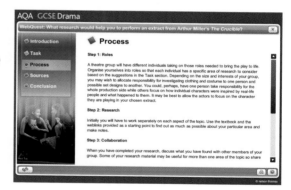

For more information and help on how to use **kerboodle** visit **www.kerboodle.com**.

How to use this book

Objectives

Look for the list of **Learning Objectives** based on the requirements of this course so you can ensure you are covering the key points.

Study tip

Don't forget to read the **Study Tips** throughout the book as well as answering **Practice Questions**.

Visit **www.nelsonthornes.com** for more information.

GCSE Systems and Control Technology

Introduction

■ What is Systems and Control Technology?

Systems and Control Technology is a very exciting, creative and interesting subject to study. It is about making products or systems that combine parts or components; these parts or components work together to control an activity. Systems always have inputs, a process and at least one output. Systems and Control Technology involves:

- understanding about materials and components
- modelling and prototyping
- developing skills that allow you to make products or systems
- learning about components that can be used as inputs, processes and outputs
- testing, investigating and evaluating products and systems
- being creative and designing new products and systems
- understanding how microcontrollers can be used to control the inputs and outputs of a system.

Our world is full of systems. As you go about your daily routines you will be influenced and affected by products that have been designed and made using Systems and Control Technology. We live in a digital age and the products around us are increasingly using microcontrollers to control the various systems within them, with many of them being made in factories using other control systems.

Your GCSE grade will be awarded as a result of completing two units of work:

Unit 1

A written examination worth 40 per cent of the total marks which will require you to apply what you have learned during the course in an examination situation.

Unit 2

A coursework project called Designing and Making Practice which involves answering a design brief and designing and making a control system; this is worth 60 per cent of the final mark.

In both your designing and making practice and the written examination, you will be assessed on how you demonstrate your knowledge, skills and understanding.

What is this book about?

You will want to achieve a good GCSE Systems and Control Technology grade. This book has been written to support your learning. You will see that the book provides you with clear, concise explanations, descriptions and examples to support you through the course and also helps prepare you for your Designing and Making Practice and the Written Examination. It has been written as a student book to support you as you go through the GCSE course.

The book is divided into two units which match the AQA GCSE Systems and Control Technology Specification:

Unit 1 Written paper

- Materials and components.
- Design and market influences.
- Processes and manufacture.

Unit 2 Designing and Making Practice

In addition to information the book will provide examples of practice questions at the end of written paper sections.

Systems and Control Technology beyond GCSE

Studying Systems and Control Technology can lead to exciting and well-paid career opportunities for skilled people to work in industry. Further courses and training are available at many different levels if you become interested in a career in this field.

In conclusion

As you focus on different topics with your teacher this book will help you gain the knowledge, understanding and skills you need to be successful. Learn as much as you can about the subject and experiment and model your ideas whenever possible.

What will you study in this section?

After completing Materials and components (Chapters 1–5) you should have a good understanding of:

materials and making

components

inputs

processes

outputs.

Materials and components

■ Introduction

Design and Technology involves the study of materials and components. The materials and components we use in Systems and Control Technology are **electronic components**, **pneumatic** and **mechanical components**, **fixings**, **mountings** and **woods**, **metals** and **plastics**. When studying Systems and Control we need to understand the properties and functions of the materials and components we use.

You will be given the opportunity to gain knowledge and understanding of the functions, working characteristics and processing techniques when designing and making Systems and Control products.

■ Materials, making and components

If the Systems and Control products you produce are to be high-quality, you will need a working understanding of materials, making processes and components. This should be sufficiently deep for you to make appropriate and reasoned choices when designing and making your control system.

There is a limited range of materials that define the areas of revision necessary for the examination, but it is expected that you will be able to show a general knowledge of the properties and characteristics of a wider range of materials through your coursework.

You will learn how common materials used in control systems can be manipulated. You will also understand what a jig and fixture is and the situations where a jig and fixture may be used. You will learn about modelling kits and software used for prototype circuits and systems, and read about their advantages for designers.

You will also need to understand different circuit prototyping methods and how to use them. It is important to know which components and fasteners are available for building functioning systems. This will help both your project work and to improve your exam technique. You must also be able to select the most suitable options from a range of materials and components and justify your choice. This applies to all parts of your solution from the main processor chip to the surface finish on the casing.

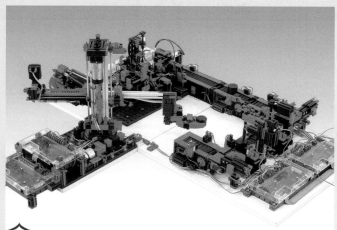

A *A model of a production line*

Input, process and output

Understanding the use of input, process and output building blocks is fundamental for this course. You should be able to create simple block diagrams for systems and learn how to apply feedback loops and where they are required.

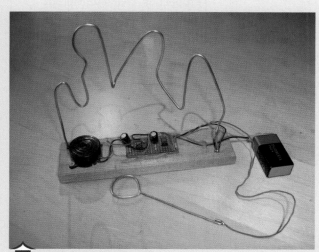

B *Steady hand game project*

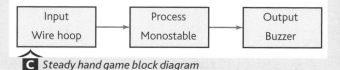

C *Steady hand game block diagram*

Photo **B** shows a simple system, a steady hand game project. It has three main parts, a loop of wire to follow, a circuit board and a buzzer. When the user accidentally touches the wire with the hoop, the buzzer sounds for half a second. The circuit that makes the buzzer stays on for half a second, regardless of how quickly the wire was touched. It is a monostable, which you will learn about in Chapter 4, and can be drawn as a block diagram. The block diagram shows the main parts of the system and how they relate to one another.

1 Materials and making processes

1.1 Materials

◼ Material choice

When designing a control system you have to select the most suitable materials based on the materials' properties, their **aesthetic** qualities, whether they can be shaped, formed or joined appropriately and effectively, their cost and availability, and increasingly other issues such as **sustainability**.

Categories of materials

Materials can be divided into different groups such as metals, timbers, plastics and ceramics. There are other ways of categorising materials: a new category of materials has recently emerged whose physical properties change according to external stimuli (temperature, magnetic fields or other inputs). These materials (referred to as **smart materials**) are made by discovering different combinations of other materials. When designing products or systems, you need to be able to explain why you have selected the materials you have.

Metals

Metals can be broken down into two sub-groups: ferrous (those containing iron) and non-ferrous (those that contain no iron). Metals can be used in their pure form or can be mixed with other metals. These mixtures are called **alloys** and are made in order to give the metals different **properties**. Steel, one of the most commonly used materials, mixes iron with a very small amount of carbon to make it harder.

Timbers

Wood comes from trees and can be placed into two main categories: softwoods and hardwoods. Softwoods generally come from quick-growing 'evergreen' trees with needle-type leaves and can be managed

A   *Some common metals and their properties*

Material	Category	Colour	Properties	Common uses
Low carbon steel	Ferrous	Grey	High **tensile** strength	Nuts and bolts, car bodies, nails
Stainless steel	Ferrous	Silver	Does not rust	Cutlery, sinks
Aluminium	Non-ferrous	Light grey	Soft, lightweight	Drinks cans, window frames, TV aerials
Copper	Non-ferrous	Reddish brown	Good conductor of electricity and heat	Tracks on PCBs, electricity cable, water pipes

as a renewable source. Hardwoods come from deciduous trees with broad leaves that are shed in winter. There are now many sheet materials made from timber products, including plywood and MDF (medium density fibreboard). These materials are generally cheaper and less likely to warp or split.

∞links

Polymorph, a new smart material, is covered on pages 12–13.

B *Some common timbers and their properties*

Material	Category	Colour	Properties	Common uses
Pine	Softwood	Generally creamy yellow	Cuts and machines well, durable	Furniture, window frames, fencing
Beech	Hardwood	Light brown	Does not split easily, fairly hard	Children's toys, workbench surfaces, kitchen utensils
Plywood	Manufactured board	Varies: creamy white to reddish brown	Reasonably stable, strong	Furniture
MDF	Manufactured board	Pale brown	Does not warp, strong	Cupboard doors, cabinets

C *Some common plastics and their properties*

Material	Category	Properties	Common uses
Acrylic	Thermoplastic	Quite hard and brittle, rigid	Safety spectacles, car rear-light units
Polypropylene	Thermoplastic	Rigid, high impact, strong	Plumbing pipes, stackable school chairs
Nylon	Thermoplastic	Hard-wearing	Gear wheels, nuts and bolts, toothbrush bristles
High density Polystyrene	Thermoplastic	Stiff, strong, tough	Egg boxes, seed trays, measuring jugs
Urea formaldehyde	Thermoset	Good thermal and electricial insulator	Electrical fittings

Plastics

This group of materials is produced largely from oil and coal, although some can be made from natural materials such as cellulose from plants. Plastics fall into two main categories: thermoplastic materials and thermosetting materials. Thermoplastic materials can be reheated and shaped many times, whereas thermosetting materials can be shaped only once.

Smart materials

This group of materials change their physical properties in response to a change in input. Piezoelectric devices can either use very small forces applied to them to produce electrical charge, or use changes to the voltage applied to them to create small movements. Shape memory alloys have been developed to remember their shape under specific conditions. An example is muscle wire – wire that contracts when heated or when a current is passed through it.

Study tip

Never refer to materials as wood, metal or plastic, etc. Learn the names and properties of a selection of each category.

Summary

Materials can be divided into different categories.

Different materials have different properties.

Designers can choose materials for their properties and for their appearance.

Manipulating materials

Making processes

When considering what you are going to make your product or system from, you need to think about how you are going to cut, shape and join the materials. The choices you make will be influenced by your experience, your knowledge of what can be done to different materials, and the equipment available to you. The main processes that can be carried out can be classified into four basic groups: cutting and removing material from the original piece; shaping by bending or forming; moulding or casting into a shape; and joining together pieces of material.

Cutting and removing material

Material can be cut and removed in a variety of ways from the three main categories available.

Common methods you might come across in school are:

- sawing to cut materials using a variety of saws
- laser cutting, using strong laser beams to burn through materials
- abrasives to wear down and smooth materials – for example, glass paper and silicon carbide paper
- drilling, using drill bits to put holes into material
- routing, removing material using a tool called a router
- planing, removing thin shavings of material with a plane
- milling, removing material with revolving cutters using milling machines
- turning, removing material by rotating the material and cutting along its length or across its end with a machine called a lathe.

Objectives

Learn how common materials can be manipulated.

Gain knowledge of how to carry out different manipulation processes.

Key terms

Forming: changing the shape of materials.

Thermoplastics: plastics that can be reshaped by reheating.

Polymorph: a material that can easily be formed or moulded at quite low temperatures.

links

See pages 14–15 for information on using a jig with a machine tool.

A *Some of these methods and the materials to which they may be applied*

Material Category	Sawing	Laser cutting	Abrasives	Drilling	Routing	Planing	Milling	Turning
Metals	X	X*	X	X			X	X
Plastics	X	X	X	X	X		X	X
Timbers	X	Some manmade boards	X	X	X	X		X

* Laser cutters used in most schools do not cut metals at present.

Shaping by bending or forming

Materials can be bent or **formed** into shapes in a variety of ways. Among the processes commonly available in schools are vacuum forming, and bending using strip heaters to shape **thermoplastics**. Vacuum forming requires a former to be made first.

B *A vacuum former, typical of the ones used in schools*

C *You may have used a strip heater like this in school*

Moulding or casting

Manipulating materials in this way requires them to be in a pliable state so they can be poured or pushed into a mould or shaped in another way. Generally this requires them to be heated. **Polymorph** is a fairly new smart material with a low melting point of 60°C. It can be reheated many times using hot water and shaped. When it cools, it becomes rigid, and cutting and removal techniques can be applied to it.

Fabricating

Materials can be joined together by various techniques including gluing (using adhesives), with mechanical fixings (e.g. rivets, nuts and bolts), by soldering or welding.

D *Adhesives*

Adhesive	Best used for joining:
PVA	timbers, some textiles
Tensol cement	acrylic
Hot melt glue	most materials but best used in the modelling stage
Epoxy resin	a wide variety of materials and can join different categories of material (e.g. metals to plastics)

> **Activity**
>
> Create a chart to show what materials and processes you would you use in making a simple protective case for a printed circuit board (PCB).

> **Summary**
>
> Materials can be manipulated by:
>
> cutting and removing
>
> bending or forming
>
> moulding or casting
>
> fabricating.

> **Study tip**
>
> When designing, give consideration not only to the materials you have available to you but also to what equipment and techniques are available to manipulate them.
>
> It is important to develop good soldering technique in order to get good electrical connections.

1.3 Jigs and fixtures

When preparing to make a product, there are a number of things that can be done to make sure that it is made correctly at the first attempt. One of the most common approaches is to use jigs and fixtures. These are simple mechanical devices that are used during manufacturing processes to help to make sure that the part is made correctly to the design.

Jigs

Jigs are made to help hold or position a part. They are used to achieve a consistent, repeated end result. If you are doing an operation once, it may not be worth making a **jig**. However, if the process needs to be repeated, a jig can save a lot of time and effort.

For example, a jig might be used if holes have to be drilled accurately and consistently in 50 PCBs. It would be designed to hold the **PCB** in the same position each time and to provide a guide for the drill bit, so that every hole was in the right place. Diagram **B** shows a suitable jig for a very simple PCB, containing six holes. The PCB is clamped into a holder by the use of a cam. The drill bit would be guided by hardened steel bushes. These are used to prevent wear and so that they can be replaced, if necessary, to extend the life of the jig.

Jigs are used with almost every manufacturing process, ranging from cutting and drilling to soldering. They can be used with hand tools, tools operated by machines and computer-controlled tools. Many jigs are designed to allow the part being worked on to be swapped over quickly and easily. A common feature of this type of jig is one or more **toggle clamps**, to hold the part in place (see Photo **C**).

Fixtures

Like a jig, a **fixture** is also a type of work-holding or positioning device. A fixture is different from a jig only in that it is normally fixed to the machine tool being used – for example, by being bolted in place – whereas jigs are normally moveable in order to line up with the tooling.

The advantages of using jigs and fixtures

Whilst these devices can help to ensure that products are made right the first time, this has to be balanced against the time needed to make them. It might not be cost-effective to make a jig or fixture

Objectives

Explain what jigs and fixtures are and how they are used.

Explain the benefits of using jigs and fixtures.

Key terms

Jig: a work-holding or positioning device that is not fixed to a machine bed.

PCB: abbreviation for printed circuit board.

Toggle clamp: a clamp that uses a locking mechanical linkage called a toggle mechanism.

Fixture: a work holding or positioning device that is fixed to a machine bed.

A *A jig is used to help fit a kitchen worktop*

if only one part or product is needed. However, there can be many advantages when making small quantities of parts.

- Jigs and fixtures reduce the amount of time needed to set up machining operations, by reducing the marking out required.
- They ensure that all of the parts are made the same.
- They reduce the amount of products that are scrapped due to being made incorrectly or need to be reprocessed.

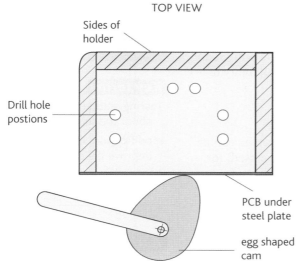

TOP VIEW

Sides of holder

Drill hole postions

PCB under steel plate

egg shaped cam

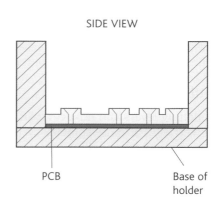

SIDE VIEW

PCB

Base of holder

B *Jig used to drill holes in PCBs*

Activities

1 Use the internet to find a range of examples of different types of jigs and fixtures.

2 Look at a product you have manufactured recently and consider the changes you would make to manufacture a batch of the product. Can you design a jig and make use of a fixture for part of the manufacturing process?

Summary

A jig is a removable device used to help hold or position a part. A fixture is a similar positioning or work-holding device that is attached to a machine base.

Jigs and fixtures are most often used when making quantities of products. They reduce the time used to set up manufacturing processes, improve consistency and repeatability and reduce the amount of scrap.

C *A toggle clamp*

∞links

Check out pages 64–65 on Mechanical systems and toggle clamps.

Study tip

You should be able to recognise situations where it would be useful to use a jig and fixture.

1.4   Modelling and prototyping

Modelling enables you to try out ideas quickly without making a whole system. This allows **testing** and **modifications** to be made. Modelling is an essential part of designing and can make savings in time and money, as well as in materials and components if the ideas don't work.

▊ Methods of modelling and prototyping

Modelling methods range from using card, glue and paper fasteners to sophisticated modelling kits or computer software. How you will model your system depends upon many factors, including the availability of modelling materials, kits and computer software.

Modelling mechanical systems

For modelling levers and linkages, stiff card and paper fasteners can be used effectively. For more complex systems, where gears are used, for example, a range of modelling kits is available. There are also software packages for modelling ideas.

Modelling and prototyping pneumatic systems

There is also software for modelling pneumatic systems. This provides an opportunity for testing ideas without having to build the system and connect to compressed air supplies.

A *Modelling mechanical systems*

Modelling and prototyping electronic systems

There are many ways to model an electronic system. Kits are available for building up electronic systems or allowing **discrete components** to be connected together.

Computer modelling

Computer software can allow electronic systems to be built and tested in a virtual environment. Advantages of using this method are:

Objectives

Understand the advantages of modelling kits and software for designers.

Understand how modelling kits and software are used for prototype circuits and systems.

Key terms

Testing: gathering information about the success of the system.

Modification: making changes to improve the system.

Discrete component: a single electronic component.

Activities

1   Using a construction kit, model mechanisms that achieve the following changes in types of motion:

  a   rotary to reciprocating

  b   rotary to linear

  c   rotary to oscillating

  d   linear to rotary.

2   Sketch your models diagrammatically.

Remember

Modelling using these methods gives you the opportunity to try out your ideas and check if they work before building more permanent systems.

⚭ links

See pages 60–61 on Outputs: types of motion.

- It is very quick.
- You can try out ideas without buying all the components.
- Component values can be altered quickly and easily.
- Simulated test equipment can be used and some results displayed on graphs.
- Errors done by trying out can be quickly rectified without component costs.

Prototype boards

Prototype boards are referred to as breadboards because they were originally made from wooden board, resembling a bread-cutting board, used with valves/tubes. Prototype boards have rows of holes connected by copper strips hidden beneath the surface. You can push the legs or pins of components into these holes to connect them without the need for soldering, but it is important you understand which ones are connected.

The process for assembling a breadboard is shown in the diagrams below.

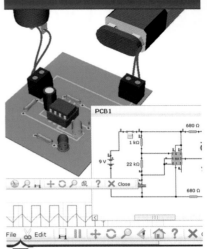

B *An example of electronic computer modelling software*

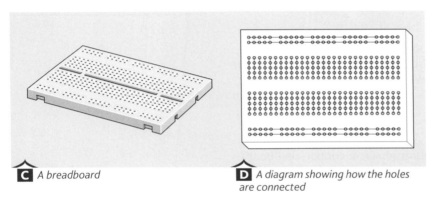

C *A breadboard*   **D** *A diagram showing how the holes are connected*

When a component or wire is pushed into a hole on the breadboard, there are more holes in that row or column providing possible connection points to other wires and/or components. This gives a quick and simple way of constructing temporary circuits. In Diagram **F**, a simple LED circuit could be built on a breadboard like this.

Study tip

Make sure you understand the advantages and disadvantages of the different prototyping methods.

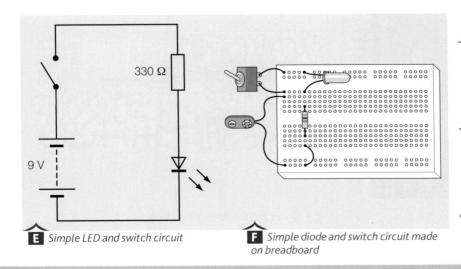

E *Simple LED and switch circuit*   **F** *Simple diode and switch circuit made on breadboard*

330 Ω

9 V

Summary

There is a range of methods available in order to model and prototype systems.

Modelling to test out ideas can demonstrate whether or not the design will work without having to make a prototype from expensive materials.

Modelling circuits enables you to check they function as intended, and allows you to alter the values of components if necessary.

1.5 Circuit construction

■ Building permanent circuits

Pages 16–17 covered modelling and prototyping electronic circuits. Breadboards are for modelling and are intended for temporary construction to test out circuit ideas, but if a more permanent solution is required for a product or system then a printed circuit board (PCB) will be required. PCBs are made on non-conductive boards covered in a thin layer of copper. Some of the copper is removed to leave tracks for connecting components.

Making PCBs

PCBs are normally made in schools by using one of the two following methods:

Etching

1. Making a mask

To **etch** a PCB, a **mask** needs to be produced. These can be made by hand, but mostly they are printed off from computer designs onto clear acetate sheets using software that allows the designer to draw out the PCB design, or software that converts a circuit model into a PCB design automatically.

2. Exposing the board

The mask is then used to expose the photosensitive copper-clad board to ultraviolet (UV) light in a UV lightbox.

3. Developing the board

The photosensitive copper-clad board is then placed in a developing solution where the protective layer that has been exposed to the UV light will be removed. It is then washed in water to stop the process by removing all traces of the developing chemical.

> ### Objectives
>
> Learn how to build circuits using a variety of methods.
>
> Know which circuit construction techniques are appropriate in different situations.

> ### Key terms
>
> **Etch**: to cut into or remove metals with strong acid.
>
> **PCB mask**: a transparent film with image of circuit placed onto it.

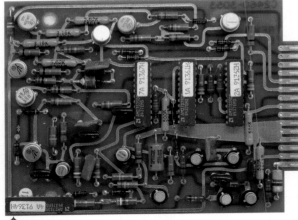

A A copper circuit board

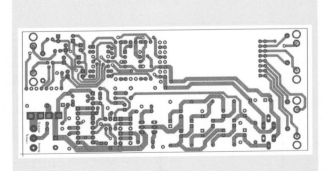

B Layout pattern for etching

4. *Etching the board*

The board is now exposed to an etching solution (ferric chloride) by dipping it in a tank or by spraying a solution onto it in a tank.

5. *Washing and drilling*

Finally the board is washed and cleaned, and is ready for drilling and soldering.

Machining PCBs

Computer-aided design (CAD) software can be used to determine the areas of copper to be removed from a board so as to leave tracking and pads where components can be placed and soldered. It is similar to the etching process in that areas of copper are removed, but it uses no chemicals and has the advantage that holes can be placed into the pads during the milling process.

C *Typical computer-controlled milling machine used in schools for PCB production*

∞ links

Compare the methods used for permanent circuits in this spread with the methods used for temporary circuits and breadboards described in the previous pages.

Activity

Make a chart to show the advantages and disadvantages of the different methods of circuit construction. Include methods for temporary circuits such as breadboards.

Study tip

- Make sure you can name the different methods of constructing circuits.
- Keep checking the circuits you construct against your circuit diagrams as carefully as possible.

Summary

There is a range of methods available to model circuits.

Modelling circuits allows you to check they do function as intended and allows you to alter the values of components if necessary.

Printed circuit boards should be used if permanent solutions are required.

1.6 Assembly

Components and fasteners

There is a not only a range of electronic components available to you but also a range of other components and fasteners that will help when fastening your PCB to your system, when placing indicator LEDs in or on your system, as well as purpose-made components to hold or fasten down motors amongst many others.

Sourcing a suitable component or fixture for your design can save time and expense in the production process, and enhance the function and appearance of your product or system. Component manufacturers produce large catalogues which will be available to you in school or college, but they also have websites that are generally easy to browse or search.

A A captive nut

B A screw and cap washer

C Students taking care while soldering

Assembling PCBs

After PCBs have been manufactured they then need to be **populated**. If a PCB has been designed using a computer software package then quite often a **real world view** can be printed off to show where components are to be placed.

Soldering

The method of making sure that you have a good electrical connection between components and copper PCB track and that they stay in place is called soldering. The copper tracks should be carefully cleaned to remove any traces of chemical or dirt or grease left on them by rubbing them with a special abrasive block designed for the purpose. Place resistors and integrated circuit holders first, then solder. More sensitive components such as transistors that may require **heat sinks** should be left until later. With the component you require to solder in place, you should begin soldering.

- Place the soldering iron so it touches the leg or pin of the component and the copper track. **Both** must be hot.
- Allow three seconds for the track and component leg/pin to heat up, and then melt a small amount of solder to onto the copper track.
- When the solder has run into a shape that resembles a small volcano, remove the soldering iron.
- You should attach a crocodile clip to the legs of heat-sensitive components to act as a heat sink and stop all the heat from the soldering iron travelling into the component.

∞links

Check out the website:
www.rapidonline.com

Activities

1   Go online to the website given in the link and find components that will do the following:
- hold PCBs in position
- hold LEDS in position
- hold motors in position.

2   Make a chart to show the order number of each of the components you find, their individual price and other information such as size of the component and component function.

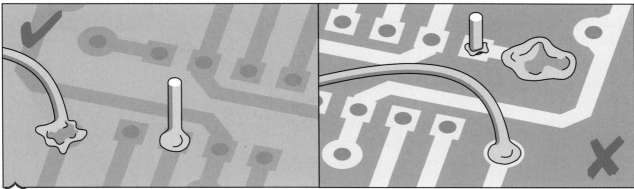

D *Good and bad soldered joints*

Summary

There is a range of components and fixtures you can use to improve your designs.

Finding suitable components and fixtures can improve the appearance and reliability of your system.

Soldering is the way that components are fixed to PCBs to allow good electrical connections.

1.7 Fault finding

If you check for quality as you make a circuit on PCB this will help eliminate faults. For example after your PCB has been made and before it has been populated there are two quality checks you can make:

■ Make a visual inspection of the copper tracks for quality and to check for any breaks.

■ Check the continuity of each track, using a multimeter.

To check for continuity, the multimeter should be switched to the ohm symbol on the dial and the probes placed either end of each track. Some meters will give an audible signal if the track is complete. Others will show low or zero resistance if complete.

If, after populating your PCB, the circuit does not function, you need to work through a series of checks and tests to find the fault.

Short circuits and dry joints

After soldering you may have one or both of these common faults, a **short circuit** and/or a **dry joint**. A visual check might help you see a short circuit, but dry joints are not so obvious.

Short circuits commonly occur where copper component pads are very close together and slightly too much solder is used, resulting in two pads that should not be joined together becoming joined by solder. The remedy is to remove some solder and make sure there is no electrical contact between the pads. A solder sucker is useful for this purpose.

Dry joints are difficult to see, but often occur where the soldering has been poor, and either the component leg or the copper track has not been heated sufficiently for the solder to stick and make good electrical contact. The remedy is to use a soldering iron to heat the joints again sufficiently to ensure the solder sticks both to the copper track and to the leg or pin of the component.

A systematic approach

The following questions might help you find the fault in your electronic system:

■ Does your PCB match the original circuit design you tested?

■ Have you placed all the components in the correct place?

■ Are the polarised components, capacitors, diodes, LEDs and ICs in the correct way round?

■ Is the battery snap/power supply connected the correct way round?

■ If a motor or other output device does not work, check if there is a voltage across it, and then check if there is sufficient current to operate it.

■ Use a meter to measure the voltage at appropriate parts of the circuit. For example, is the supply reaching the supply rails and key components like ICs?

Key term

Short circuit: a type of electrical circuit failure.

Dry joints: soldered joints that do not provide an electrical connection.

A A multimeter

B A solder sucker

Measuring voltage and current with a multimeter

In the circuit shown below the probes from a meter can be placed as shown to measure voltage. The multimeter dial must be turned to the correct setting to measure voltage.

To measure current the multimeter needs to be placed in the circuit in series with a component; again the dial must be placed in the correct position to measure current.

Activities

1. Make up the monostable circuit shown. Connect the multimeter in parallel to the capacitor and watch the reading on the meter when the capacitor charges.

2. Make your own tick chart to list the checks you can make to test and fault-find your next circuit.

links

Refer back to Diagram **D** on page 21.

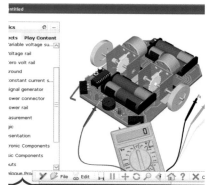

C *A multimeter is used to check the supply voltage*

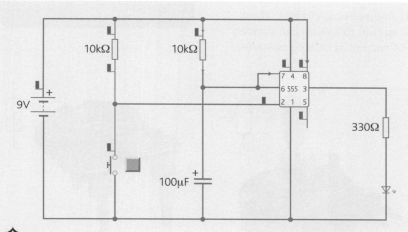

D *Monostable circuit*

Summary

You need to check for quality continually when making a PCB circuit.

There are two common soldering faults: dry joints and short circuits.

If a circuit does not work, check the voltage with a meter at key points in the circuit.

Study tip

Continually check for quality in each step of production from mask to each soldered joint. Look for breaks in tracks in PCBs, incorrect circuit layouts, components in the wrong place and polarised components in the wrong way round.

2.1   Components

Electronic components

There are many different types of components used in electronic systems for a variety reasons. Some components are used to control the flow of current in a circuit and to divide voltage, some store electrical charge, others only allow current to flow freely in one direction and not the other, amongst many other things.

Resistors

Resistors are very common components in electronic circuits. They are used to limit the flow of current in a circuit. Resistors can divide voltage but not set it. Limiting the flow of current is called resistance and is measured in ohms (Ω).

A *Fixed resistors*   **B** *Potentiometer*   **C** *Preset*

Capacitors

Capacitors can perform a variety of functions in electronic circuits but they are most commonly used to control the delay in timing circuits. They work by storing electrical charge. When they are connected to a power supply and a current flows into them, they charge up. If you then connect together the two leads of the capacitor in a circuit, the capacitor discharges its stored electrical energy. The time it takes for a capacitor to charge depends upon two things:

- the size of the capacitor – the larger it is, the longer it takes to charge
- the speed of the charging current – dependent on the resistance the current is flowing through.

Capacitance is measured in farads (F), and because the farad is a very large unit it is more commonly measured in microfarads, millionths of a farad (μF).

links

For more information about potential dividers, check out pages 40–41.

D *Non-polarised capacitor*

Diodes

A diode only lets current through in one direction and therefore is like a one-way street for electrical current. They are commonly used to protect transistors in switching circuits when magnetic fields collapse, for example in a coil-operated relay.

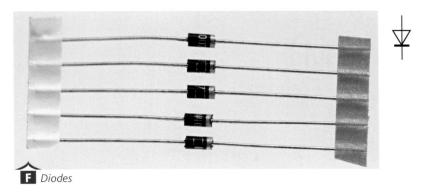

F *Diodes*

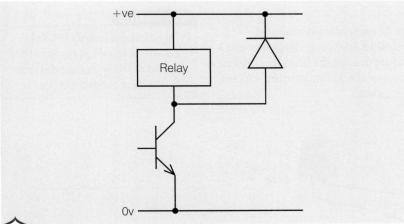

G *Circuit diagram of a relay, transistor and protective diode*

Protective diode

When an electro-magnetic device like a solenoid or the relay in the diagram de-energises a large back **electromotive force (EMF)** is created when the magnetic field collapses. The back EMF can rise to a much greater level than the supply voltage and permanently damage the transistor; the protective diode prevents this happening.

2.2 Component symbols

Why use circuit symbols?

Symbols are used when drawing diagrams because they simplify circuit drawing. Look at the pictures of the components below.

Objectives

Understand the function of component symbols.

Be able to name different components from their circuit symbols.

Know how to use component symbols when drawing circuit diagrams.

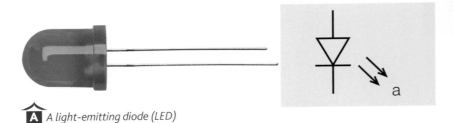

A *A light-emitting diode (LED)*

 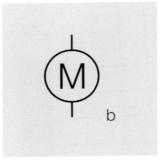

B *A motor*

The symbols for these components are drawn as shown above.

Circuit diagrams

It would be very difficult to draw pictures of each component and which other components they are connected to so circuit diagrams are used. Circuit diagrams use circuit symbols with straight lines to connect them, these lines represent the wire or copper tracks on **printed circuit boards (PCBs)** that connect them to each other.

Key terms

Printed circuit boards (PCBs): Boards onto which components are soldered with copper tracks connecting them together.

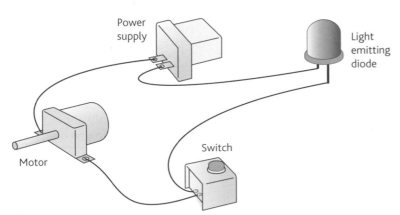

C *A group of components connected together by wire to form a circuit*

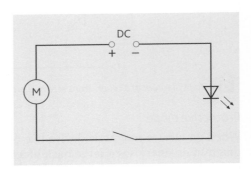

D *A circuit diagram*

Constructing circuit diagrams

When drawing circuit diagrams designers always try to construct them so they can be read from left to right. Power supply rails are drawn at the top and bottom of the circuit with the positive supply rail running along the top and the negative rail along the bottom. Input components are placed on the left of the diagram and it progresses through process components to output components on the right.

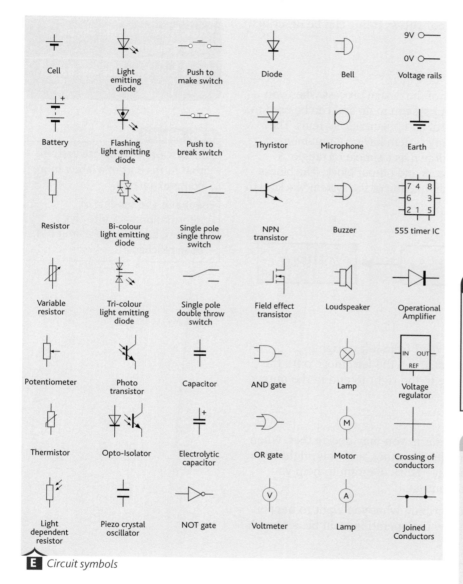

E *Circuit symbols*

Study tip

- Learn a few circuit symbols at a time and test yourself or get others to test you.
- Make sure you know every circuit symbol and know what the actual component looks like.

Activities

1. Make cards for all the component symbols and quiz yourself and others.

2. Get a number of electronic components from your teacher and place them next to the card you have made with the components symbol on it.

Summary

Circuit symbols are used to simplify circuit diagrams.

Circuit diagrams are generally read from left to right.

Every electronic component has its own symbol.

∞ links

See pages 34–36 for information about input components.

2.3 Electronic building blocks

What is a system?

A system is a collection of parts that exists to perform a function. In Systems and Control Technology we break a system down into smaller blocks; each block describes a function in the control system. The operation of any system or its individual **sub-systems** can be described by using a block diagram.

Block diagrams

Block diagrams can be used at the start of the design process when you have an idea of what you would like your system to do and can be used to describe electronic, electrical, mechanical and pneumatic systems. They are made from a series of blocks with any system having a minimum of three blocks. A control system block is drawn as a simple rectangle. A system is always made of an input, process and output block. The blocks in the system are linked by a signal. The function of the system blocks is to change the signal in some way.

A *An input, process, output block diagram*

Block diagrams are made so they are read from left to right.
The input block or blocks start the system on the left, move onto the process block or blocks, then continue to the right to end in the output block or blocks.

Ice alarm system

If you were wanting to design an ice alarm you may decide that, when it gets cold, you want to hear an audible warning, and you might begin your system design by drawing a simple block diagram to help you think in terms of input, process and output.

A block diagram is very useful in describing what you want to happen in a system and in helping decide how the operations will be achieved.

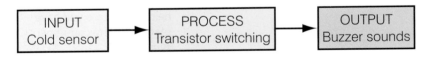

B *A simple block diagram for an ice alarm system*

An automatic curtain-closing system

The example opposite shows the beginning of a design for a block diagram of a system that automatically opens and closes curtains when it becomes dark or light.

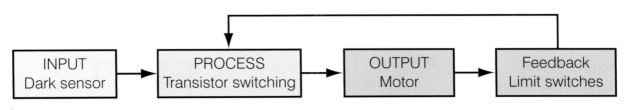

C *A block diagram for an automatic curtain-opening system using feedback*

This system uses **feedback** in order to stop the motor when the curtains fully close. Thinking about your system in this way may also help you see other processes that need to happen for your system to operate effectively. In this system a relay was added to work as an interface. This enables the motor to turn in the opposite direction to open the circuits when it gets light again by using the switch contacts on the relay as a reversing circuit.

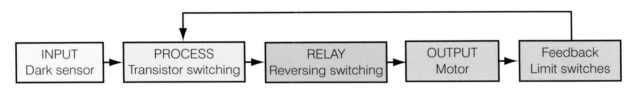

D *A development of the automatic curtain-opening system with a block added for a relay reversing system*

Activities

1   Analyse and draw systems block diagrams for a set of automatic opening doors and an automatic security lamp.

2   Draw a block diagram for an automatic greenhouse window-opening system and show any feedback loops you think may be necessary.

Study tip

When researching and analysing existing systems in project work always break them down into input, process, output, and ask yourself if feedback is used.

Summary

A block diagram is a way of describing what happens in a system.

A systems block diagram always has a minimum of one input, process and output block.

Feedback is where information is sent back into the system.

◌◌ links

See pages 78–79 for more information on types of system.

See pages 100–107 on flow diagrams.

Using formulae

Your knowledge of maths and applying formulae can help you make design solutions to problems.

Ohm's law

Ohm's law describes the relationship between **voltage**, **current** and **resistance**. It states that the current passing through a resistor is proportional to the voltage across it. Basically this means if you double the voltage across a resistor you will double the current flowing through it; should you treble the voltage you will treble the current and so on. Also, if you double the size of the resistance and leave the voltage constant you will halve the current, treble the resistance and you will reduce the current to a third and so on. This is known as Ohm's law.

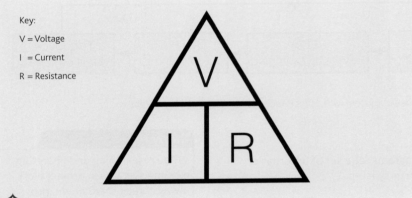

Key:

V = Voltage

I = Current

R = Resistance

A This triangle is often used as an aid to remembering Ohm's law

If we know any two of these we can work out the other using these formulae:

$V = I \times R$

$I = V \div R$

$R = V \div I$

For example, if we connect a 12 ohm resistance in a circuit with a voltage across it supplied by a 6-volt battery we could work out current flow as follows:

$I = V \div R$

$I = 6 \text{ volts} \div 12 \text{ ohms}$

$= 0.5 \text{ amps}$

Objectives

Understand how to use formulae to calculate values.

Understand when to apply the formulae given in the specification.

Series resistance

If two or more resistors are connected together in series as in the diagram below we can work out the resistance by adding the values together.

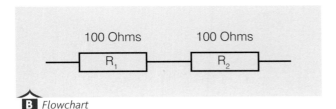

100 Ohms   100 Ohms

B *Flowchart*

This can be written as a formula

$R_T = R_1 + R_2$

(Total)   (resistor values)

$R_T = R_1 + R_2$

Rt = 100 + 100

Rt = 200 ohms

Potential dividers

We can work out voltage at certain key points in a circuit where a **potential divider** has been used using this formula.

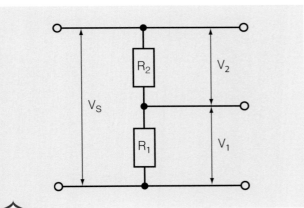

C *Circuit diagram*

links

See pages 40–41 for more information on potential dividers.

Key terms

Potential divider: a resistor connected in series between positive (+ve) and 0v supply of a circuit, used to divide the supply voltage.

If there were a supply voltage of 9v and R_1 and R_2 were both 100 ohms and we needed to know voltage V1 we would use the following formulae:

$$\text{Voltage 1} = \frac{R1}{R1 + R2} \times \text{Supply Voltage}$$

$$\text{Voltage 1} = \frac{100}{100 + 100} \times 9$$

$$\text{Voltage 1} = \frac{1}{2} \times 9$$

Voltage 1 = 4.5 volts

$$\text{Voltage 2} = \frac{R2}{R1 + R2} \times \text{Supply Voltage}$$

$$\text{Voltage 2} = \frac{100}{100 + 100} \times 9$$

$$\text{Voltage 2} = \frac{1}{2} \times 9$$

Voltage 2 = 4.5 volts

Gear ratios

The ratio between the size of **driver** gear and the size of **driven** gear is called the gear ratio. This also determines the relative speed of the gears and can be referred to as the **velocity ratio**.

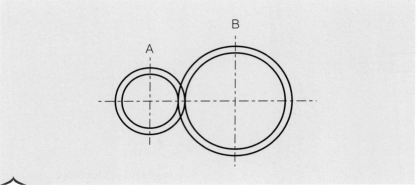

D *Gear ratio*

We can use this formula to work out the gear ratio where a driver gear (A) has 20 teeth and a driven gear (B) has 40 teeth.

$$\text{Gear ratio} = \frac{40}{20} = \frac{4}{2} = \frac{2}{1} = 2{:}1 \text{ or } 2$$

Velocity ratios

When working with belt and pulley systems we can apply a very similar formula to calculating gear ratios. This is in order to work out the relative speed of each of the two pulleys.

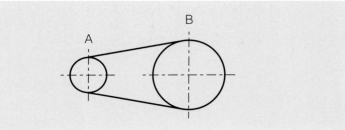

E *Velocity ratio*

This formula can be used to work out the velocity ratio where a driver pulley (A) has diameter of 20mm and a driven pulley (B) has a diameter of 40mm.

$$\text{Velocity ratio} = \frac{\text{Diameter of driven pulley}}{\text{Diameter of driver pulley}}$$

$$\text{Velocity ratio} = \frac{40}{20} = \frac{4}{2} = \frac{2}{1} = 2{:}1 \text{ or } 2$$

Output speed

If we know the speed a driver gear or a driver pulley is turning at we can calculate output speed using this formula:

$$\text{Output speed} = \frac{\text{Input speed}}{\text{Gear/Velocity ratio}}$$

∞ **links**

See pages 68–69 for more information about gears.

We can use this formula to calculate the output speed of the pulley in the previous example if the driver pulley was turning at 200 **rpm**.

$$\text{Output speed} = \frac{\text{Input speed}}{\text{Gear/Velocity ratio}}$$

$$\text{Output speed} = \frac{200}{2} = 100 \text{ rpm}$$

Mechanical advantage

In a mechanism where a smaller effort can be used to move a bigger load, the mechanism is said to give a mechanical advantage. Mechanical advantage can be calculated using the following formula:

$$\text{Mechanical advantage} = \frac{\text{Load}}{\text{Effort}}$$

In a mechanism, if a load of 400N were moved by applying an effort of 100N we could work out the mechanical advantage given by the mechanism.

$$\text{Mechanical advantage} = \frac{\text{Load}}{\text{Effort}}$$

$$\text{Mechanical advantage} = \frac{400}{100} = \frac{4}{1} = 4{:}1 \text{ or } 4$$

links

See pages 64–65 for more information on mechanical advantage.

Key terms

rpm: the speed something turns measured in revolutions per minute.

N: abbreviation for newtons, the units used to measure force.

Activity

1   Use the formulae on this page to perform the following calculations.

a   Work out the resistance if a current of 0.03 amps is measured in a circuit with a supply voltage of 9 volts.

b   Work out the total resistance if a 1k resistor is placed in series with a 680 ohm resistor.

c   Work out what voltage you would get across R_1 in a potential divider circuit where R_1 and R_2 are 2k and 4k respectively, in a circuit with a supply voltage of 0v.

d   Work out the gear ratio in a simple gear train where the driver gear has 25 teeth and the driven gear has 100 teeth.

e   Work out for the above gear train the output speed if the driver gear is turning at 160 rpm.

f   Work out the mechanical advantage in a system where an effort of 20N is used to move a load of 120N.

Study tip

■ Remember any formulae you need will be printed in the final examination paper.

■ Remember to include the formula you use, your working out, and the units used.

Summary

A formula is a key to solving a mathematical problem.

By substituting values into formulae and calculating the answer, you can solve design problems.

3.1 Inputs: sensors

Using the potential divider circuit to create sensors

To make a **sensor** circuit, a resistor that can change its resistance depending upon its surroundings can be chosen for one of the resistors.

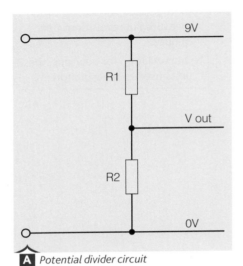

A *Potential divider circuit*

Objectives

Understand that potential dividers can be used to create sensors.

Learn about different types of input sensor.

Learn how to calibrate a sensor circuit.

Key terms

Sensor: a transducer that converts one type of energy to another, e.g. light to electricity.

Calibrate: to check, adjust or determine by comparison with a standard.

⚮links

To find out more about potential dividers, go to pages 40–41.

B   *Sensors and the components to build them*

Light or dark sensor	Light-dependent resistor (LDR)	
Hot or cold sensor	Thermistor	
Wet or dry sensor	Two probes	

A variable resistor should be used for the other resistor so that the midpoint voltage can be adjusted.

The six common sensor circuits

There are six common sensor circuits, diagrams for which are given on page 36. Please note that in the light/dark sensor circuit, it is the position of the LDR that determines whether the potential divider midpoint goes high when it is light or when it is dark.

This also applies to the position of the thermistor in the hot/cold sensor, and the position of the probes in the wet/dry sensor.

C *Examples of uses for the six common sensor circuits*

Light sensor	Operate window blinds when the sun shines
Dark sensor	Turn on street lights when it gets dark
Hot sensor	Open a vent when a greenhouse gets too hot
Cold sensor	Operate a heater when it gets too cold
Wet sensor	Operate a pump if water is found in the bottom of a boat
Dry sensor	Water a plant if the soil is too dry

Calibrating the sensors using the variable resistor

The sensor circuit can be adjusted or **calibrated** by altering the variable resistor. This will ensure that the sensor circuit will give a suitable voltage at the midpoint, at the correct light or heat level, and so on.

The output from the sensor could be compared with a known accurate reading, from a thermometer for example, to ensure that the sensor circuit was working accurately.

Activities

1 Build one of the sensor circuits on breadboard. Change it to work in the opposite way.

2 Identify and sketch products that use each of the six sensor circuits shown on page 36.

D *Dark sensors are used to turn on street lights when it gets dark*

kerboodle

The six common sensor circuits

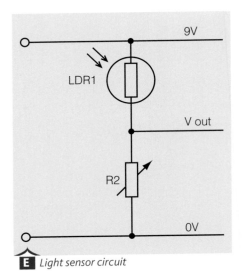

E *Light sensor circuit*

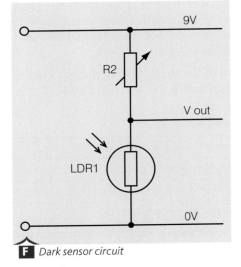

F *Dark sensor circuit*

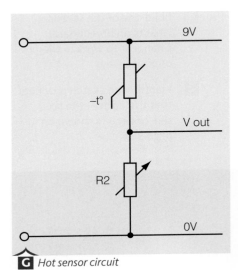

G *Hot sensor circuit*

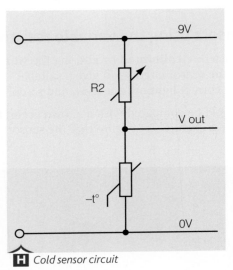

H *Cold sensor circuit*

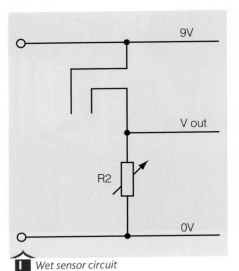

I *Wet sensor circuit*

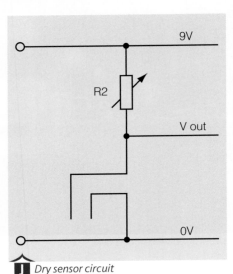

J *Dry sensor circuit*

Summary

An LDR can be used to sense light and dark.

A thermistor can be used to sense hot and cold.

Metal probes can be used to sense wet and dry.

A sensor circuit can be calibrated to give a specific voltage at a certain condition.

3.2   Inputs: switches

Push switch

The push **switch** is the simplest form of switch. It can be made to connect the circuit when pressed 'push to make' or break the circuit when pushed 'push to break'.

Objectives

Recognise the different types of switch.

Know when to use the different switch types.

Key terms

Switch: a device to break or connect an electric circuit.

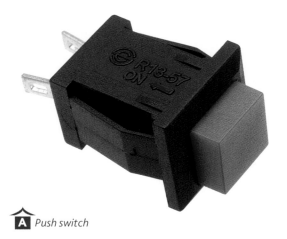

A *Push switch*

B *Push switch circuit symbols*

Push to make   Push to break

The 'push to make' and 'push to break' switches look the same, so care must be taken to select the correct one. An example use for a 'push to make' switch is in a house doorbell circuit; when pressed the bell rings. An example use for a 'push to break' switch would be in a window switch for a burglar alarm. If the window were opened the circuit would be broken and an alarm could be made to sound. The advantage is that the alarm would sound if the burglar cut the wires to the switch.

Types of switch

Switches are classed by how many poles and how many throws they have. The pole is how many circuits the switch can make or break, the throw is the number of contacts that the pole can be connected to.

In the circuit below an SPST switch has been used as the ON/OFF switch and an SPDT is used to decide which bulb to light.

C *Different types of switch*

	SPST	Single pole single throw
	SPDT	Single pole double throw
	DPST	Double pole single throw
	DPDT	Double pole double throw

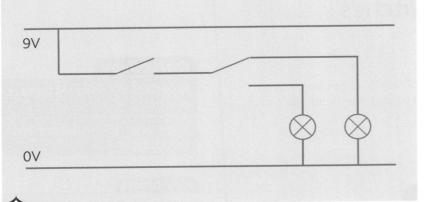

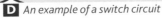

D *An example of a switch circuit*

Forms of switch

Micro-switch

The micro-switch is a popular switch as it is small and can be easily mounted. They are found inside complex electro-mechanical systems – for example, washing machines and printers. They are available in many sizes and values, and there are many types of lever to operate them.

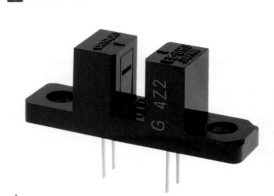

E *Micro-switch*

Opto-switch

An opto-switch consists of an infrared LED and a phototransistor combined in a single package. The slotted opto-switch shown in the photo detects when an object is in the slot. It can be used to sense when a disc or track is in a certain position. They are used in printers and robotics.

F *Opto-switch*

Reed switch

The reed switch is an SPST switch encased in a glass tube. The switch is activated when a magnet is placed near to it. The advantage is that the switch can be operated without needing to be touched. They are used in alarm circuits and proximity detectors in automated systems.

G *Reed switch*

Tilt switch

The tilt switch is an SPST switch that is operated by tilting it to a certain angle. They can be used to sense the angle of a part of a system or sound an alarm if a device tilts too far.

Key switch

The key switch is operated, as it names implies, with a key. They are used to reset alarm systems.

Reversing motor circuit with DPDT switch

A DPDT switch can be used to reverse an electric motor. It does this by reversing the polarity of the power supply. Note that the SPST switch is required to turn the motor ON and OFF.

H *Key Switch*

I *Tilt Switch*

∞ links

To find out more about switches search these suppliers' websites:

Rapid Electronics:
www.rapidonline.com

RS Components:
uk.rs-online.com

Technology Supplies:
www.technologysupplies.co.uk

Activities

1 Find a use for each of the switches shown on these pages.

2 Use the internet or component catalogues to find prices for each of the common switches.

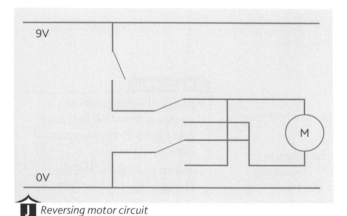

J *Reversing motor circuit*

Study tip

- Make sure that you understand the terms and diagrams for SPST, SPDT, DPST, DPDT.

- You should be able to recognise and suggest an application for the micro-switch, opto-switch, reed switch, key switch and tilt switch.

- Make sure you can show how a DPDT switch can be used to reverse a motor.

Summary

Push switches can be 'push to make' or 'push to break'.

Switches can be SPST, SPDT, DPST, DPDT.

Forms of switch include; micro-switch, opto-switch, reed switch, tilt switch, key switch.

Motors can be reversed with a DPDT switch.

Ohm's law

As explained in Chapter 2, Ohm's law describes the relationship between **voltage**, **current** and **resistance**.

It states that voltage (V) = current (I) × resistance (R)

If any two of the above are known, the third can be calculated.

In the simple circuit on the right:

- Voltage = 10V
- Current = 10mA
- Resistance = 1K

$V = IR$

$V = 10mA \times 1K$

$10V = 0.01mA \times 1,000\Omega$

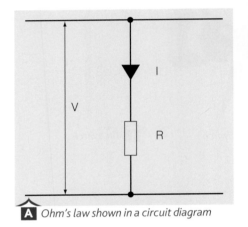

A Ohm's law shown in a circuit diagram

The potential divider circuit

The potential divider circuit is used to create sensor circuits which can be used as inputs into systems.

In its simplest form the potential divider comprises two resistors in series between the power rails (9V and 0V) of the circuit. The supply voltage (**electrical potential**) is split (divided) between the two resistors and the voltage at the midpoint (V out) can be at any point between 0 volts and the supply voltage.

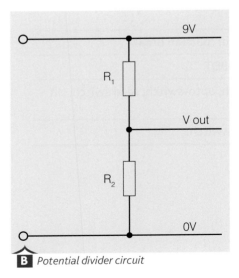

B Potential divider circuit

Using Ohm's law to calculate V out for potential divider

In the potential divider circuit below, R_1 is 10K and R_2 is 10K.

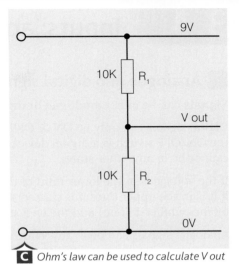

C *Ohm's law can be used to calculate V out*

D *Ohm's law formulae*

Calculate the total resistance	$Rt = R_1 + R_2$	20K = 10K + 10K
Calculate the total current	$I_t = V_t/R_t$	0.45 mA = 9V/20K
Calculate the voltage across R2	$V_2 = I_t R_2$	4.5V = 0.45 mA x 10K

Potential divider with similar resistors

As the example in Diagram **C** shows, if the two resistors are of equal resistance, the supply voltage is halved.

Note: when building a potential divider it is important to use high value (above 10K) resistors as lower values would short out the power supply.

Using the potential divider circuit to create sensors

To make a sensor circuit, a resistor that can change its resistance dependent on surroundings is used for one of the resistors, and a variable resistor for the other.

In the light sensor circuit shown on page 36, for example, the resistance of the LDR will change as the light level changes. This alters the voltage (potential) across it, thereby changing the voltage at the midpoint (V out).

As the light level increases, the resistance of the LDR reduces, as does the voltage across it. This increases the voltage across the variable resistor, V out. A process block connected to this would see the voltage rise and make the relevant decision.

⬭links

To find out more about sensors, refer back to pages 34–35.

Activities

1   Using Ohms Law, calculate voltage (V) when the current (I) is 10mA and the resistance (R) is 1K.

2   Explain, using Ohm's law, how a dark sensor could be built using an LDR and a variable resistor.

Study tip

■ You should be able to calculate voltage, current and resistance using Ohm's law. The formula will be given to you in the exam.

■ Make sure that you can explain the operation of the potential divider sensor circuit.

■ Check that you can calculate V out for a sensor circuit.

Summary

V = IR

The potential divider gives a variable voltage that is dependent upon the two resistor values.

If the two resistors are the same, V out is half the supply voltage.

3.4   Inputs: analogue and digital

■ Analogue and digital signals

Signals can be either **analogue** or **digital**.

Digital signals can only be ON or OFF. An example would be a torch; the ON/OFF switch is a digital device and can only be ON or OFF. It cannot be in any other **state**.

If the voltage is in the lower third of the range it is classed as OFF and if it is in the upper third it is classed as ON. The ON state is represented by the number 1 (one) and the OFF state by a 0 (zero).

Analogue signals can be at any value between a minimum and maximum. They have an infinite number of possible values in between. An example would be a mercury thermometer. Assuming that it is designed to show any temperature between −20°C and 120°C, there are an infinite number of values that the mercury can show, depending upon how far up the thermometer it is.

Analogue   Digital

A *Analogue and digital signal waveforms*

Objectives

Understand the difference between analogue and digital signals.

Understand the use of digital and analogue input sensors with PICs.

Key terms

Signal: a message between devices.

Analogue: any value within a range.

Digital: either ON or OFF; 1 or 0.

State: the condition of an input or output.

Peripheral interface controller (PIC): a microcontroller that can be programmed to react to various input sensors and give a variety of outputs.

■ Analogue and digital inputs

Input devices can be classed as either digital or analogue depending on whether they have a set number of states (usually just two) or an infinite number of states.

■ Analogue and digital inputs into a PIC

A **peripheral interface controller (PIC)** is a microcontroller that can be programmed by the user to react to various input sensors and give a variety of outputs. PICs can be purchased with either analogue or digital inputs or, more usually, a combination of the two. The designer should choose a PIC with the correct number and type of inputs. In a washing machine controller for example, the door switch would be connected to a digital input and the water temperature sensor would be connected to an analogue input. The door can be either open or closed and the water temperature can be at any value between 5°C and 60°C.

The circuits in Diagram **E** show how to connect an analogue and digital input to a PIC.

B *The ON/OFF switch on this torch is a digital device*

C *A mercury thermometer is an analogue device*

D *Analogue and digital input devices*

Digital input devices	Analogue input devices
Slide switch	Light sensor (LDR)
Rotary switch	Temperature sensor (thermistor)
Reed switch	Water sensor (probes)
Touch switch	Position sensor (variable resistor)
Tilt switch	

In the first circuit, the switch varies the voltage at the midpoint connected to the digital input. The digital input circuit needs the 10K resistor to pull the input down to 0V when the switch is open. Without the 10K resistor, the input would not be connected to anything when the switch was open and could pick up stray signals. If its value were less than 10K it would short out the power supply when the switch was closed.

The analogue input circuit uses a thermistor to sense the temperature. Its resistance changes depending upon its temperature. This resistance change varies the voltage at the midpoint connected to the analogue input.

Depending upon where the sensor is to be used it requires calibrating to ensure that the voltage is within a suitable range for the PIC. To calibrate the sensor, the 22K variable resistor is adjusted.

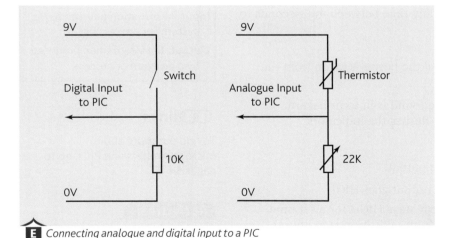

 Connecting analogue and digital input to a PIC

Summary

Signals can be either analogue or digital.

Digital signals can only be ON or OFF.

Analogue signals have an infinite number of possible values.

PICs can have analogue and digital inputs.

Analogue sensors need to be calibrated.

Study tip

■ Ensure that you can identify analogue and digital inputs.

■ You should be able to select a suitable input for a given purpose.

■ Make sure you understand the components used in the common sensor circuits.

4.1 Time delay

Time delays

A time delay is used in a **process** to add time. Without them the **program** would run too quickly and it would be unusable. For example, in a program in a washing machine, a time delay is used to ensure that the clothes are washed for 30 minutes.

Astable, monostable

Two of the common forms of time delay are the astable and monostable. These can be made from various components or programmed into a PIC. It is important to understand what they can do.

Astable

- The astable has no **input** and one **output**.
- When powered up the output goes HIGH (this can be any time between milliseconds to hours).
- It then goes LOW (this can also be any time between milliseconds to hours).
- It then goes HIGH.

This process continues indefinitely until the power is taken from the astable.

An example of where an astable can be found is in a car hazard warning system. When switched on, it flashes the indicators.

Monostable

The monostable has one input and one output.

- When the input is made HIGH, the output goes HIGH.
- When the input goes LOW, the output stays HIGH for a set time period (this can be set to any time from milliseconds to hours).
- The output then goes LOW and waits to be triggered again.

An example of where a monostable can be found is in a security light. When the sensor senses body heat, the light comes ON. The light then stays ON for a set time, even if the person runs away. Another example would be in a burglar alarm: once triggered, the alarm stays on for 30 minutes even if the burglar closes the door or window that triggered the alarm sensor.

Monostable and astable waveforms

The inputs and outputs can be shown on a graph of voltage against time.

Objectives

Understand time delays in systems.

Be able to create time delays using electronic circuits.

Be able to create time delays by programming software.

Key terms

Process: a computing operation.

Program: a set of coded instructions that enables a computer to perform a desired sequence of operations.

Input: information put into a system for processing.

Output: the information produced by a program or process.

⊂⊃ links

To find out more about microcontrollers and PICs, go to pages 54–55.

Activities

1. Sketch three products that use a monostable and three that use an astable.

2. Calculate the component values for a 555 monostable circuit to give a 10 second delay.

⊂⊃ links

To find out how to build an astable from a 555 timer, go to pages 46–47.

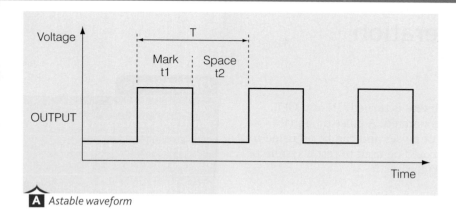

A *Astable waveform*

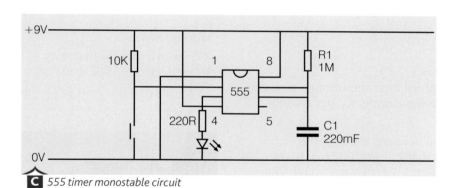

B *Monostable waveform*

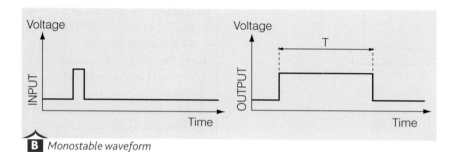

C *555 timer monostable circuit*

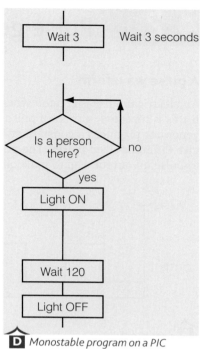

D *Monostable program on a PIC*

Formulae to change the timing period

Monostable formula

The value of R1 and C1 can be selected to set the time that the monostable stays high after being triggered.

$$T = 1.1 \times R1 \times C1 \qquad T = 1.1 \times 1M \times 220uF \qquad T = 242 \text{ seconds}$$

Time delay using a PIC

A time delay can easily be achieved with a WAIT command, as shown on the flowchart. This flowchart is part of the process from a security light. When the sensor senses a person, the light comes on for two minutes.

Summary

The astable gives a pulsed output.

The monostable gives a HIGH output for a set period.

The astable and monostable can be built from a 555 timer.

The time periods can be set by choosing the correct values for the resistors and capacitor.

The WAIT command gives a time delay in a PIC.

4.2 Pulse generation

A pulse waveform

An electrical pulse is created when a device is turned ON and OFF. If this is repeated, a train of pulses is generated. A pulse or signal generator produces a constant stream of pulses and can be adjusted to give different-sized pulses. The pulses could be used to test a system or to pulse an output, e.g. to make an LED flash.

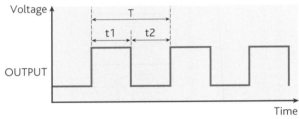

 **A** *A pulse waveform*

Frequency and mark/space ratio

The **frequency** of a waveform is the number of complete waveforms (T) that occur in a second. In the above diagram a complete waveform is labelled T.

The formula to calculate frequency is:

$$\text{Frequency} = \frac{1}{T}$$

The unit of frequency is **hertz** (Hz), named after the German physicist Heinrich Hertz. 1Hz means 1 complete waveform once per second; 100Hz is 100 per second, etc.

The **mark/space waveform** is the ratio between the time the waveform is high (mark) and the time that it is low (space). In the above waveform the mark space ratio is 1:1, i.e. they are equal. The mark/space ratio can be adjusted in a pulse generator to give the required output.

Producing a pulse train with an astable

An astable is a pulse generator as it generates a stream of pulses. If the astable is set to pulse slowly it can be used to operate something every few hours, for example a pet feeder or a plant-watering device. If the astable is made to pulse very quickly (10KHz) it can be connected to a piezo transducer or loudspeaker to make a sound.

Astable made with a 555 timer

The astable can be made with a 555 timer.

Formulae to change the timing period of the astable

The value of the two resistors and the capacitor in the astable circuit can be selected to give the required frequency of output.

 **B** *An astable set to pulse slowly can ensure your pet gets fed*

∞links

To find out how to build a monostable from a 555 timer go to pages 44–45.

Objectives

Understand what is meant by pulse generation and frequency.

Understand how frequency is measured.

Learn how to set a frequency.

Key terms

Frequency: the number of repetitions per unit time of a complete waveform.

Hertz: the SI unit of frequency. One hertz is one cycle per second.

Mark/space waveform: the ratio between the high and low part of a waveform.

Worked example

If T = 20mS then the frequency = 1/T or 1/20mS or 1/ 20 × 10⁻³ = 50Hz

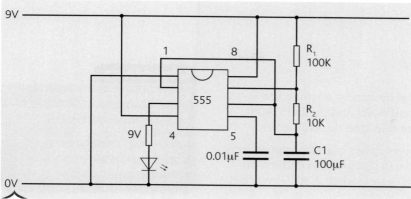

C *555 timer astable circuit*

$t_1 = 0.7 \times (R_1 + R_2) \times C_1$

$t_2 = 0.7 \times R_2 \times C_1$

$T = t_1 + t_2$

$t_1 = 0.7 \times (10K + 100K) \times 100uF$

$t_2 = 0.7 \times 100K \times 100uF$

$t_1 = 7.7$ seconds

$t_2 = 7.0$ seconds

$T = 14.7$ seconds

The frequency (F) can also be calculated by using the formula on page 46 or the following formula:

$$F = \frac{1.44}{(R1 + 2R2)\, C1}$$

$$F = \frac{1.44}{(10K + 200K) \times 100uF}$$

$F = 0.068$ Hz

Pulse generation using a PIC

The diagram on the right shows how a PIC can be used as a pulse generator. When the program reaches the loop it goes around the loop turning the LED ON and then OFF repeatedly. The Wait commands are used to determine the frequency and mark/space ratio.

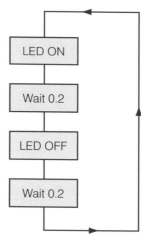

D *A PIC used as a pulse generator*

Activities

1   Build a pulse generator with a 555 timer and adjust its frequency by changing components.

2   Connect a pulse generator to a speaker and work out which frequencies are required to play a simple tune.

3   Draw the flowchart for a simple traffic light.

4   List systems (products) that have a pulse generator in them.

Study tip

- Make sure you understand the terms frequency, hertz and mark/space ratio.
- Be able to label the mark and space of a waveform.
- Make sure that you can give examples of where an astable is used.
- Check that you can calculate the frequency of a waveform.

Summary

The astable gives a pulsed output.

Frequency is measured in hertz.

Pulses can be any speed from very slow to very fast.

A PIC can be used to generate a pulsed wave form.

Switch circuits

In the simple **switch** circuit, the circuit is broken by opening the switch. When the switch is closed, current flows and the bulb lights. When the switch is open, current cannot flow and the bulb is off.

In the series switch circuit there are two switches one after the other. For the bulb to light BOTH switches have to be closed so current can flow. This behaviour is similar to an AND gate: switch A AND switch B need to be closed for the bulb to light.

In the parallel switch circuit there are two switches side by side. For the bulb to light EITHER switch can be closed so current can flow. This behaviour is similar to an OR gate: switch A OR switch B can be closed for the bulb to light.

Objectives

Understand that different circuits can be used to switch on a variety of outputs.

Learn about transistors operating as amplifiers and switches.

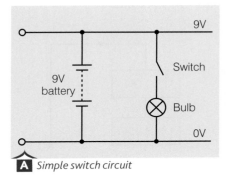

A Simple switch circuit

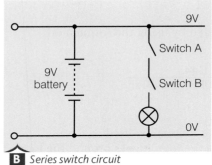

B Series switch circuit

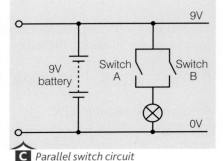

C Parallel switch circuit

Explaining the transistor circuit

In the Diagram **D**, there is nothing connected to the current-limiting resistor that is connected to the base of the **transistor**. The voltage at the base is therefore 0V. The transistor is OFF, current cannot flow from the collector to the emitter and the bulb is OFF.

The current-limiting resistor is required to protect the transistor. If it were not there, too much current would flow and the transistor would fail.

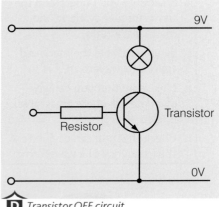

D Transistor OFF circuit

Key terms

Switch: a device to break or connect an electric circuit.

Transistor: an electronic component that can be used as a switch or amplifier.

Threshold voltage: the voltage that a transistor switches ON, usually 0.6V.

Insulator: a substance that does not conduct electricity, e.g. rubber.

Conductor: a substance that conducts electricity, e.g. copper.

In the circuit on the right, the current-limiting resistor is connected to the 9V rail; the voltage at the base of the transistor is therefore 9V. The transistor is ON, current can flow from the collector to the emitter and the bulb is ON.

In this example, the voltage at the base was either 9V when the lamp was ON or 0V when it was OFF. The transistor does not require 9V to switch ON; a voltage above 0.6V will switch it ON. This is called the **threshold voltage** and is the voltage that the transistor changes from an **insulator** to a **conductor**.

Transistor used as a switch

In this circuit a push to make switch has been added to connect the transistor base to 9V. When the switch is not pressed, the base is at 0V; this is below 0.6V so the transistor is acting as an insulator and the bulb is OFF.

When the switch is pressed, the base is at 9V; this is above 0.6V so the transistor is acting as a conductor and the bulb is ON.

(Please note: circuit has little use and is only used to demonstrate transistor switching.)

Transistor used as an amplifier

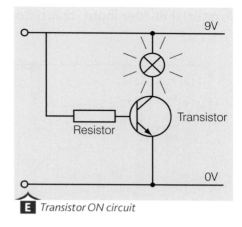
E *Transistor ON circuit*

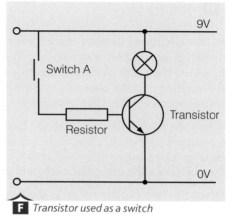
F *Transistor used as a switch*

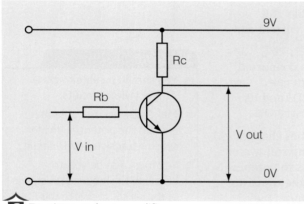
G *Transistor used as an amplifier*

The transistor can be used as an amplifier by producing a large change in the collector current for a small change in base current. This is called current amplification.

The transistor can also be used to amplify voltage because the resistor Rc reacts to these large changes in collector current which, in turn, results in large variations in the output voltage (V out).

Amplifier Gain can be calculated with the following formula:

$$Av = \frac{\text{Change in input voltage}}{\text{Change in output voltage}}$$

Potential divider input, transistor process, LED output

⊂⊃links

To find out more about how sensors can be used go to pages 40–41.

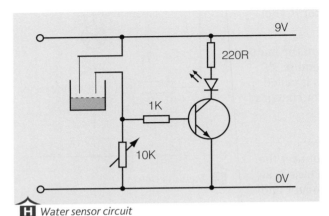

H Water sensor circuit

In this circuit we have an input, process and output. The potential divider is acting as a water sensor. The 10K variable resistor is used to adjust the point that the transistor turns on. The 1K resistor limits the current into the base of the transistor to protect it, and the 220R resistor protects the LED by limiting the current through it.

When the two probes are **dry**, the base of the transistor is not connected to 9V but is connected to 0V through the variable 10K resistor. The transistor is OFF and acts as an insulator. No current can flow between the collector and emitter and the LED is OFF.

When the two probes are **wet**, the base of the transistor is connected to 9V. The transistor is ON and acts as a conductor. Current can flow between the collector and emitter, connecting the LED (and its protective resistor) between the 9V and 0V rail; the LED turns ON.

This happens because the resistance between the two probes changes depending on what is between them, e.g. air, dry soil, damp soil, water. This change in resistance causes a change in voltage across the probes due to Ohm's law. This changes the voltage at the base of the transistor and can make it higher than the threshold voltage which turns the transistor ON.

Activities

1. Draw the circuit for the interior light and door switches for a 4-door car.

2. Draw the simplified circuit for the door switch, ON switch and heater in a microwave oven.

3. Describe products that could use the water sensor circuit in Diagram **H**.

4. Build the simple sensor transistor circuit in Diagram **F** to light an LED when the probes are placed in water.

5. Draw a circuit that operates a buzzer if the temperature goes above a certain level – for example, a freezer alarm.

Summary

Switches can be connected in series and parallel.

The transistor can be used as an electronic switch.

The transistor can be used as an amplifier.

4.4   Logic

Logic

In this form, **logic** refers to the rules that determine a process. The process does not need to be electronic or in a computer. For example, to be able to listen to a piece of music, I must download a music track or buy the CD.

Here, we are interested in electronic logic; this determines the output of a **digital** system. Digital inputs and outputs can only be **ON or OFF**, which is represented as **1 or 0**. The process building blocks that we are going to consider on this page are called **logic gates**. Logic gates always behave in the same way, i.e. they would always give the same output with the same set of inputs. They are called gates because they work in a similar way to a garden gate – they can allow or stop the flow of something through them. Logic gates can be purchased in multiples on DIL chips **intergrated circuit (ICs)** or represented on a program in a PIC (microcontroller).

The common logic gates: NOT, AND, OR

The following logic gates are recognised as the simplest to understand and use. They have internationally recognised symbols as shown below. The inputs and outputs are given letters to identify them. These can change and do not have to be as shown below.

NOT gate

The NOT gate is the simplest logic gate. It has one input and one output; the output is NOT the input.

When the input is ON the output is OFF. When the input is OFF the output is ON.

OR gate

The simple OR gate has two inputs and one output; OR gates can have multiple inputs.

The output is ON when any of the inputs are ON.

When the input A OR input B are ON the output is ON.

All of the inputs have to be OFF for the output to be OFF.

AND gate

The simple AND gate has two inputs and one output; AND gates can have multiple inputs.

The output is only ON when all of the inputs are ON.

When the input A AND input B are ON the output is ON.

If any of the inputs is OFF, the output is OFF.

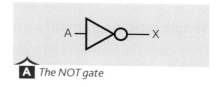

A *The NOT gate*

B *The OR gate*

C *The AND gate*

Truth tables for logic gates

A **truth table** can be drawn to show all of the possible inputs to a logic gate or logic circuit and the expected output(s). On truth tables 1 and 0 are used instead of ON and OFF (1 = ON, 0 = OFF).

> **Key terms**
>
> **Truth table**: a table that shows a system's inputs, outputs, and all of the possible states.

D *NOT truth table*

A	X
0	1
1	0

The input is A and the output is X.

X = NOT A

If A OR B are 1, then X is 1.

E *OR truth table*

A	B	X
0	0	0
0	1	1
1	0	1
1	1	1

The inputs are A and B, the output is X.

X = A OR B

F *AND truth table*

A	B	X
0	0	0
0	1	0
1	0	0
1	1	1

The inputs are A and B, the output is X.

X = A AND B

If A AND B are 1 then X is 1.

Simple logic diagram for a greenhouse heater

Logic gates can be combined together to form a logic diagram. The outputs of previous gates form the inputs to the next gates. Complex systems with many inputs and outputs can be represented this way.

In Diagram **H**, which shows a circuit for a simple greenhouse heater, there are three inputs (the manual override, the timer and the temperature sensor). Each input can be ON (1) or OFF (0). There is one output, the heater, which can also either be ON (1) or OFF (0):

G

Input	ON	OFF
Manual override	Heat	Don't heat
Timer	Heat	Don't heat
Temperature sensor	Too hot	Too cold

The heater should be ON when the manual override OR the timer is ON, AND the temperature sensor is NOT too hot.

links

To find out more about digital signals go to pages 42–43.

To find out more about how switches can be used as logic gates, go to pages 48–50.

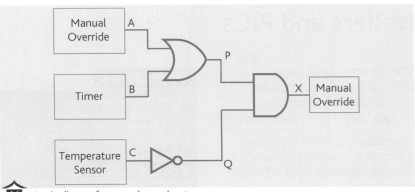

H Logic digram for greenhouse heater

Activities

1 Draw the truth tables for a 3-input OR and 3-input AND gate.

2 Sketch a microwave oven. Draw the logic diagram to make sure that the door is locked before it works.

3 Sketch a simple washing machine. Draw the logic diagram.

![A greenhouse]

I A greenhouse

Study tip

■ Ensure that you can recognise the NOT, AND, OR logic symbols and truth tables.

■ Be able to complete truth tables for a simple logic circuit.

■ Make sure you can design a logic circuit for a given scenario.

Truth table for the greenhouse heater logic diagram

A truth table can be created for the greenhouse heater logic diagram.

First add the 3 inputs, A, B and C. Ensure that all of the possible input combinations are shown.

Add the 2 intermediate points P and Q, P = A OR B, Q = NOT C Complete their states.

Finally, add the output X, X = P AND Q

J A truth table for the greenhouse heater

A	B	C	P	Q	X
0	0	0	0	1	0
0	0	1	0	0	0
0	1	0	1	1	1
0	1	1	1	0	0
1	0	0	1	1	1
1	0	1	1	0	0
1	1	0	1	1	1
1	1	1	1	0	0

As can be seen the heater only comes ON on three occasions, when either the timer OR the manual override is ON, AND it is too cold.

Summary

Logic gates are process blocks, and can be electronic or used in programs.

NOT gives output 1 when input is 0.

OR gives output 1 when any input is 1.

AND gives output 1 when all inputs are 1.

Gates can be combined in multiple input and output logic diagrams.

A **microcontroller** is a computer on a chip. A common form of microcontroller used in schools is a peripheral interface controller (PIC). Microcontrollers are found in many household items; any electrical product more complex than a torch probably contains one!

A PIC can be programmed to behave in a certain way depending upon its surroundings. The program can be written as a series of commands or instructions or as a flowchart. The PIC flowchart programming software converts the flowchart to a series of commands that the PIC understands.

PICs are purchased in **dual in line (DIL)** chips and can be bought with various numbers of legs: 8, 14, 18, 20, 28 and 40 pin.

Objectives

Learn the functions of microcontroller ports.

Understand how to select the appropriate microcontroller for different functions.

Learn about control software packages used to control the inputs and outputs of a microcontroller.

Types of port, input, output, dual

Some of the legs on the PIC are for power and control. The rest (ports) allow signals into and out of the PIC.

Input ports are used to enter signals into the PIC. These could come from switches or sensors. Input ports can be digital or analogue. Output ports are used to send signals out from the PIC – for example, to LEDs, motors or speakers.

Dual ports can be either input or output ports. They are controlled by the program in the PIC and give more flexibility.

PICAXE–14M

```
           +V  [ 1        14 ]  0V
      Serial In  [ 2        13 ]  Output 0 / Serial Out / Infraout
   ADC / Input 4  [ 3        12 ]  Output 1
  Infrain / Input 3  [ 4        11 ]  Output 2
       Input 2  [ 5        10 ]  Output 3
       Input 1  [ 6         9 ]  Output 4
   ADC 0 / Input 0  [ 7         8 ]  Output 5
```

A *PICAXE–14M pin layout*

Ports can be either digital or analogue. Digital are far more common: if they are not labelled as analogue, assume that they will be digital. The user must decide which type would be best for the PIC's application.

This PIC has 5 inputs and 6 outputs. Note that the first input/output is numbered 0.

Picaxe and logicator software

PICs can be programmed easily with either PICAXE or Logicator software. PICAXE software uses a series of **BASIC** commands; Logicator software uses a flowchart. The examples in Table **C** and Diagram **D** would make an LED connected to output 0 flash.

B *PICAXE-14M PIC*

C *PICAXE BASIC commands*

PICAXE	Logicator
main	label a program 'main'
high 0	switch on 0
wait 1	wait 1 second
low 0	switch off 0
wait 1	wait 1 second
go to main	go to the start

Key terms

Microcontroller: a type of microprocessor, a computer on a chip.

Dual in line (DIL): a type of chip consisting of two lines of legs 0.1 inches apart.

BASIC: a simple programming language used in schools and colleges.

There isn't a flash command so the LED is turned ON and OFF repeatedly. The wait commands are required to slow the program down. There is no stop command; as the above program has a loop that returns the flow to the start, this program would continue to run until power was removed from the PIC.

Simple flowchart commands: start, decision, process, wait, stop

Start box

All flowcharts should start with this box. It shows where the program will start from.

Decision box

The decision box is used to branch the software. They are used to check the inputs and direct the program in different directions, depending on the input's state; for example, if it is dark, turn on the street light.

The outputs should be labelled with the correct values.

Process box

The process box is used to set the ports and outputs. In this example it makes output '0' high.

Wait box

The wait box adds a delay to the program. Without them, the program would run so quickly, the outputs may not show and it would finish milliseconds after it started.

Stop box

This shows that the program has reached the end. Not all programs need one as they can loop back to the start.

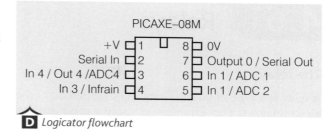

D Logicator flowchart

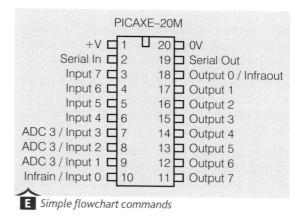

E Simple flowchart commands

∞ links

This is only a small subset of the available commands. See these websites for more information.

To find out more about PICAXE software: **www.rev-ed.co.uk/picaxe**

To find out more about Logicator software: **www.economatics-education.co.uk**

Summary

PICs can have various combinations of inputs and outputs.

PIC inputs can be digital or analogue.

PICs can be programmed in BASIC or a flowchart.

DECISION boxes allow the program to change depending upon the inputs.

WAIT boxes allow the program to be slowed down to give the required outputs.

Activities

1 Write a BASIC program for a simple 3-LED traffic light. Where are there WAIT commands in the program?

2 Design a flowchart for a simple 3-LED traffic light.

3 Design a flowchart for a railway level crossing barrier. You will need to sketch the barrier and track first. The inputs and outputs are shown below.

Input	Train Sensor North	1 Train present	0 No Train
Input	Train Sensor South	1 Train present	0 No Train
Output	Traffic Warning Lights	1 ON	0 OFF
Output	Barrier Motor Control	1 Up	0 Down

4 What other sensors/outputs could be added to the railway level crossing barrier system to improve it?

4.6 Relays

The relay

A relay is a type of switch that can be turned on and off by an electromagnet. The advantage of using a relay is that the two circuits are kept apart and are only joined by a mechanical armature. This allows a low current sensor/control circuit to operate a much larger current.

Relays create an **interface** between the process and output of a system. In this example the interface is used to convert low current signals so that they can control high current circuits.

As an example, the cables for the starter motor circuit for a car engine are very thick and need to be as short as possible between the starter motor and the car battery. A relay is used to control this circuit from the car dashboard.

The relay operates by connecting the low current so that current flows through the relay coil; this creates an electro-magnetic field. The field attracts an iron **armature**; this pushes the switch contacts together, switching the circuit ON. When the current is switched off, a spring returns the armature and the contacts open again, switching the circuit OFF.

In Diagram **D**, the electromagnet and the switch contacts can be seen. The diagram shows how the armature pivots when attracted to the electromagnet. This operates the switch contacts.

B Relay symbol

C A car starter motor uses a relay

A A relay

The relay symbol shows the electromagnet coil as a rectangle and the switch contacts. The contacts do not have to be drawn near the coil but must be labelled so it is clear which coil operates them.

Relays can have many switch contacts depending upon the need of the circuit. In this book and in the exam, very simple relay circuits are shown.

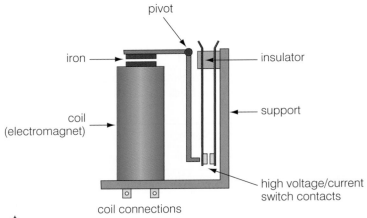

D Relay schematic diagram

The protection diode

When a relay is used as an interface between low current electronic circuits there is a danger that the relay could damage the electronic components. This is due to the electromagnetic field that is generated in the relay coil dissipating back into the electronic circuit when the relay is turned off (referred to as back EMF). To prevent damage to the circuit, a diode is placed across the relay coil. The diode must be placed in the opposite direction to the circuit's normal current flow so that it does not short out the transistor in normal operation.

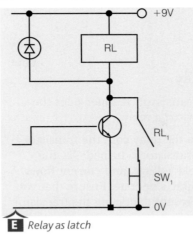

E *Relay as latch*

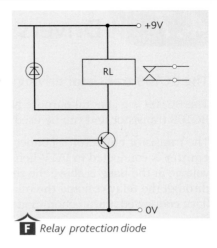

F *Relay protection diode*

The latch

A **latch** is a circuit that stays ON even if the trigger signal is turned OFF. An example of their use is in alarm circuits where they ensure that the bell continues to ring even if the burglar closes the window, steps off the pressure switch, etc.

The relay as latch

The following circuit shows the relay used as a latch. Once the relay is operated, one of the relay's contacts (RL1) allows current to flow to the coil, so even if the transistor that operated the relay turns OFF, the relay remains ON. To turn the relay OFF, the push switch (SW1) must be used to break the circuit. The relay, now without power, turns OFF.

Study tip

- Ensure that you can explain a relay circuit.
- Be able to give the advantages of using a relay.
- Make sure you know how to add a protection diode and explain its function.
- Make sure you can also complete or explain the relay latch circuit.

Activity

1 Look at the circuit diagram for a car and identify the relays and relay contacts. These can be found in workshop manuals or on the web.

  a   List the products where relays are used.

  b   Investigate where relays are used to switch very high voltage circuits.

  c   Give one disadvantage for using a relay in an alarm system. How could this be improved?

links

To find out more about switch contacts that can be operated by a relay see pages 37–39.

To find out more about relays search these suppliers' websites:

Rapid Electronics: **www.rapidonline.com**

RS Components: uk.rs-online.com

Technology Supplies: **www.technologysupplies.co.uk**

Summary

Relays are interface devices.

Relays can allow small voltage/current devices to switch large voltage/current devices.

Relays require a protection diode to prevent the back EMF damaging the electronic components.

Relays can be used as a simple latch.

4.7 Drivers

The BC109 transistor driving a relay

The BC109 is a general purpose **NPN transistor**. It supersedes the BC108 transistor and can be used to switch a relay.

The transistor **collector** is connected to the relay coil; the transistor **emitter** is connected to 0V. When the transistor is turned ON, the voltage at the **base** is above the threshold voltage, and current flows through the relay coil and the relay contacts operate. Ensure that you have connected a protection diode to prevent damaging the transistor.

Objectives

Learn about the uses of transistors and relays to switch on outputs.

Learn about different types of transistor.

Key terms

NPN transistor: a type of bipolar transistor.

Base, collector, emitter: the three legs of an NPN transistor.

Field effect transistor (FET): a type of unipolar transistor.

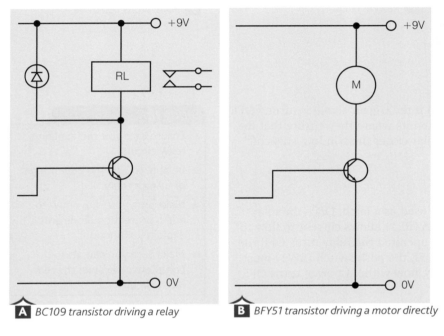

A BC109 transistor driving a relay

B BFY51 transistor driving a motor directly

The BFY51 transistor driving a motor directly

The BFY51 is a high power NPN transistor. It can be used to switch a motor directly.

In a similar way to the BC109 circuit, the transistor collector is connected to the motor; the transistor emitter is connected to 0V. When the transistor is turned ON, the voltage at the base is above the threshold voltage, and current flows through the motor and it operates.

H bridge motor control

Transistors can be used to reverse the direction of an electric motor. In the H bridge circuit, four switches or transistors are used.

In this circuit:

- If no switches are operated the motor will be off.
- If S1 and S4 are operated, the motor will spin.
- If S2 and S3 are operated, the motor will spin in the opposite direction.

links

To find out more about transistors as a switch see pages 48–50.

To find out more about relays see pages 56–57.

To find out more about transistors, search these suppliers' websites:

Rapid Electronics: **www.rapidonline.com**

RS Components **uk.rs-online.com**

Technology Supplies: **www.technologysupplies.co.uk**.

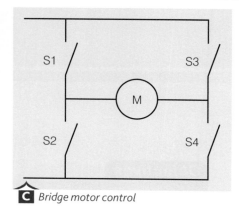

C *Bridge motor control*

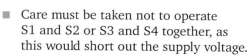

D *A field effect transistor (FET) operating a solenoid*

- Care must be taken not to operate S1 and S2 or S3 and S4 together, as this would short out the supply voltage.

The H Bridge motor control circuit is used in robotics to control the many motors from the main processor.

A MOSFET driving a solenoid

A metal oxide semiconductor, **field effect transistor (FET)** can handle higher currents than the NPN transistors and is therefore better suited to switching high current devices.

A solenoid is a coil of wire wrapped around a soft iron core. It takes a high current and can become warm to the touch.

Common transistor types

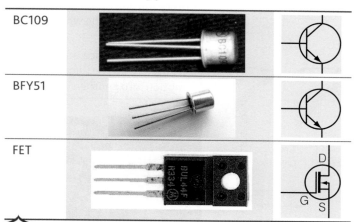

E *Table of the common transistors*

Summary

NPN transistors can be used to operate low current devices, e.g. LEDs.

The H bridge circuit allows transistors to control the direction of a motor.

MOSFET transistors can switch high current devices, e.g. solenoids.

Activities

1. Use manufacturers' catalogues and websites to find the prices of the above transistors.

2. Build a simple transistor H bridge circuit that allows a PIC to control a motor.

Study tip
- Ensure that you can design a transistor driver circuit.
- Be able to select the correct transistor for the specified use.

5.1 Motion

Types of motion

There are four types of motions. **Linear motion** is motion in a straight line, for example of a train moving along its track or an apple falling to the ground. **Reciprocating motion** is motion from side to side or up and down. **Oscillating motion** is a waving sideways movement. **Rotary motion** is movement in a circle, going round and round.

A Linear motion

B Linear motion

C Reciprocating motion

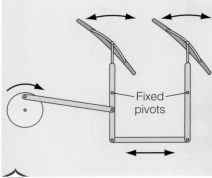
Fixed pivots
D Oscillating motion

E Oscillating motion

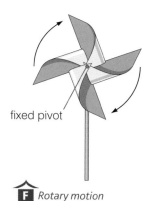

F *Rotary motion*

G *Pedalling a bicycle is an example of rotary motion*

Changing motion

Motions can be changed from one form to another, and various **mechanisms** are available to achieve this. A rack and pinion can convert linear motion into rotary motion and vice versa. You can see this mechanism on a steering rack or a pillar drill.

A crank turns rotary motion into reciprocating motion. A treadle linkage turns rotary motion into oscillating motion.

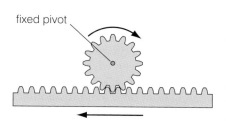

H *Rack and pinion mechanism*

⬭ links

For more information about mechanical systems see pages 64–65 and for information about gears see pages 66–67.

Check out the website: **www.flying-pig.co.uk/mechanisms** for more information about motion and mechanisms.

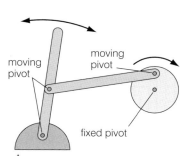

I *A crank can convert rotary motion into reciprocating motion*

J *A crank and slider mechanism changes rotary motion into reciprocating motion.*

Summary

There are four types of motion: linear, reciprocating, oscillating and rotary.

These can be changed from one form to another by different mechanisms.

Light-emitting diodes

A **light-emitting diode (LED)** is a semiconductor diode that gives out light in one direction when an electric current is applied in one direction. LEDs have many applications and are increasingly used instead of low voltage lamps. LEDs are available in many different colours and sizes, and are usually used in conjunction with a series resistor, which stabilises the current. LEDs can produce more light than a low voltage lamp given the same current. They are very small in size, have a low failure rate, light up quickly and have an extremely long lifetime. They contain fewer toxic elements than a normal incandescent lamp, and therefore are easy to recycle. Generally they are less expensive and connect easily to a printed circuit board.

Initially LEDs were used simply as power indicators. Increasingly, their applications are becoming more varied, particularly when used in numbers as an array. The range of colours produced by LEDs depends on the chemical composition of the semi-conductor used. Even white light can be produced using the primary colours green, red and blue. LEDs can even have their own function, flashing or changing from one colour to another.

Low voltage lamps

Also referred to as incandescent lamps, **low voltage lamps** consist of a very thin filament, usually made of tungsten, through which an electric current is passed. This filament is encased in a glass bulb containing

Objectives

Learn about different optoelectronic outputs and devices.

Understand the advantages and disadvantages of different optoelectronic outputs in different situations.

Key terms

Light-emitting diode (LED): a semiconductor that gives out light.

Low voltage lamp/bulb: an electric light in which a filament is heated.

Liquid crystal display (LCD): a low-powered flat panel display.

Seven segment display: seven LEDs positioned to form a figure of eight.

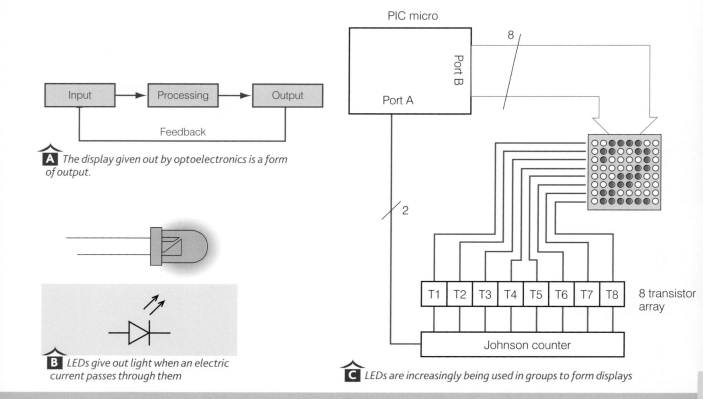

A The display given out by optoelectronics is a form of output.

B LEDs give out light when an electric current passes through them

C LEDs are increasingly being used in groups to form displays

an inert gas. They are available in many voltages from 1.5V to 500V; however, they are being rapidly replaced by other forms of lighting, mainly due to their poor energy efficiency (they give out much more heat than light). Unlike LEDs, they have a relatively short life and poor shock resistance. They need to have their own holder and are generally more expensive than LEDs.

■ Liquid crystal display (LCD)

A **liquid crystal display (LCD)** is a very thin display panel made up of a number of colour or monochrome pixels arranged in front of polarising filters and an electronic light source. They are used in a number of products from calculators to televisions. Compared to conventional displays, they are small and thin, very light, and use less energy. However, their viewing angle is quite small, they are sensitive to touch and are relatively expensive.

■ Seven segment display

Seven segment displays have a range of applications such as clocks, microwaves and alarms. They consist of seven LEDs (eight if you need a decimal point), which can be illuminated to form all of the decimal characters from nought to nine.

D *Seven segment digital displays are seen in many applications*

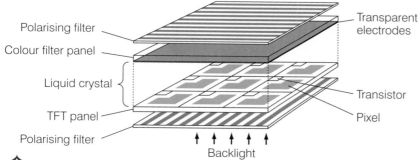

E *Liquid crystal displays are used in modern television screens*

Study tip

Make sure that you know about the advantages and disadvantages of a lamp compared to an LED. Also, remember which LEDs to light when producing 0 to 9 on a seven segment display and suggest an optoelectonic for a given application.

Summary

Light-emitting diodes (LEDs) can be found in a whole range of colours and sizes, and be grouped to form arrays.

Liquid crystal displays (LCDs) are used to convey information and pictures.

Seven segment displays are used to display decimal numbers.

◯◯ links

Take a look at the following websites:

www.play-hookey.com/digital/experiments

www.caves.org.uk/led

www.litewave.co.uk

Levers

The simplest of all machines, **levers** consist of a uniform bar, a **load**, an **effort** and a pivot called a **fulcrum.** There are three different classes of lever depending on where you position the load, effort and fulcrum. Levers are used every day in objects such as pliers, crowbars, spanners and switches. Classes one and two multiply effort, giving a **mechanical advantage (MA)**.

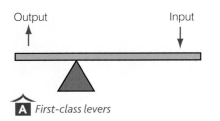

A *First-class levers*

B *A seesaw*

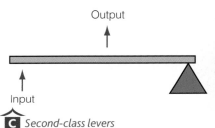

C *Second-class levers*

D *Wheelbarrows have second-class levers*

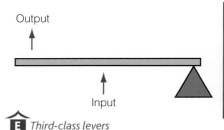

E *Third-class levers*

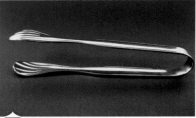

F *Sugar tongs are an example of third-class levers*

Mechanical advantage (MA)

This is the ratio of the load compared to the effort. In most cases, the MA is greater than one; however, in some cases the mechanism may be designed to gain speed or change direction and the MA could be less than one. To calculate mechanical advantage, divide the load by the effort.

For example, you could use a second-class lever (a wheelbarrow) to lift a load of 300N with an effort of 50N.

Linkages

Linkages are used extensively in mechanical systems. They allow motion to be transferred from one place to another. They can be used to change the size of the force, the direction of movement or make things move in a special way. In addition, one linkage can be attached to another linkage to make a particular type of movement.

Reverse motion linkages.

Linkages work using two types of pivot or fulcrum, a fixed pivot and a movable pivot. The magnitude or speed of the motion is governed by the positions of pivots and lengths of the component levers.

Parallel linkages

These can be found in mechanisms like scissorjacks, lazy tongs and toolboxes. They are based on a parallelogram, and the parts are designed to move in unison.

Toggle clamps

These are used extensively in the school workshop and are useful for producing locking mechanisms. A good example in the home is a louvre window catch.

Treadle linkage

Treadle linkage is a simple mechanism that changes motion from one form to another (rotary to oscillating). An electric motor produces rotary motion, but when converted by a treadle linkage, this causes car windscreen wipers to oscillate.

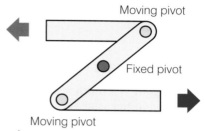
G *Reverse linkage*

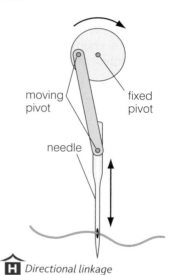
H *Directional linkage*

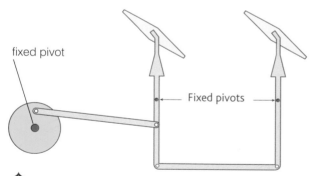
I *Diagram of a set of windscreen wipers*

J *Windscreen wipers on an aeroplane window*

Activities

1. Give three examples of each of the three classes of lever.

2. Use the notes and sketches to explain how a reverse motion linkage works.

3. How many levers and linkages can you find on a bicycle? Can you name these mechanisms? Use diagrams to illustrate your answer.

Summary

Levers are simple machines which come in three classes depending on the relative positions of the fulcrum, effort and load.

Mechanical advantage can be calculated by dividing the load by the effort.

Linkages allow motion to be transferred from one place to another.

Spur gears

A spur **gear** is a toothed wheel used to transmit motion. One gear driving another gear is called a **simple gear train**. The diagram below shows a simple gear train consisting of two gears, one called the driver and the other called the driven. The two gear axles turn in opposite directions unless you place an **idler gear** in between. By changing their size (i.e. the number of teeth) you can alter speed and power, thereby changing the **gear ratio**. The size of the gear ratio indicates how much the mechanism changes speed and power. You can calculate the gear ratio (sometimes called the velocity ratio) by using the formula:

Gear Ratio (VR) = number of teeth on the driven gear/number of teeth on the driver gear.

$$= 40/8 = 5/1 = 5:1$$

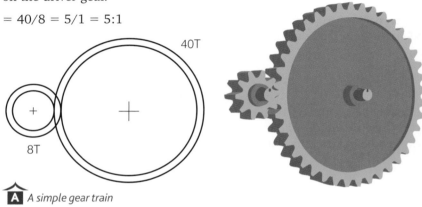

A *A simple gear train*

Gears are very difficult to sketch, so they are often represented as the symbol, shown above.

Gear trains

If you use pairs of gears mounted on the same axle or shaft this is called a **compound gear train**.

If you need large increases or reductions in speed or power it may be necessary to use a compound gear train, as in the diagram below. Compound gear trains have a range of applications from children's toys to ocean liners and give large increases and decreases in speed and power.

The easiest way of calculating the gear ratio of a compound gear train is to consider each of the pair of simple gear trains, calculate the gear ratio of each, and then multiply the gear ratios together to find a final gear ratio. This process is shown in Diagram **C**.

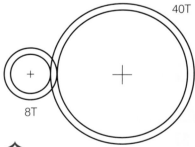

B *This is the symbol for a gear train*

Total gear ratio = GR1 x GR2 = 2:1 x 3:1 = 6:1

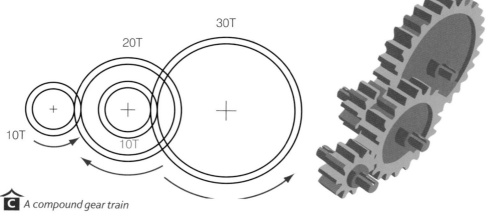

C *A compound gear train*

This would give you a total gear ratio (velocity ratio) of 6:1. In other words, you would reduce the final output speed approximately six times and increase the final output power by a factor of six.

Now you can work out changes in speed, because this gear ratio means that every time the driver gear turns around six times the last output gear will turn around once. This will reduce the overall speed by a multiple of six and therefore the output speed will be 40rpm, given an input speed of 240rpm.

Activities

1   Use a construction kit to model a compound gear train. (Your school is likely to have some such kit – if in doubt ask your teacher.) Calculate its gear ratio.

2   Find a mechanism at home or in the workshop which uses gears. This is quite difficult nowadays as most utensils have the gears enclosed for health and safety reasons, but an egg whisk with gears (not a balloon egg whisk) or a hand-operated drill would be ideal. Draw the gear train and explain how it works.

3   Using the gear symbol draw a compound gear with a velocity ratio/ gear ratio of 12:1.

∞links

See this website:
www.howstuffworks.com/gear.htm for explanations of geared machines.

Or this website:
www.dynamicscience.com.au/ tester/solutions/hydraulicus/ gears1compound.htm is useful if you find it helpful to watch animations showing gears working.

Summary

A gear is a toothed wheel used to transmit motion.

Simple gears can be meshed together to form compound gears or gear trains.

The gear ratio tells you how much the simple machine changes speed and power.

Gear trains involve a pair (or several pairs) of gears linked together, sometimes on the same axle.

Study tip

Make sure you know what a gear is, how to describe the different types and how to calculate the gear ratio.

In addition to the gears described on the previous pages, there is a range of other specialist gears which can be used in different applications. The formula for calculating velocity ratio, speed and power is the same for most of these gears.

Worm gears

If you require a larger reduction in speed, and thereby a large increase in power, you could use a worm gear. A worm gear is very like a screw thread, i.e. one continuous spiral. This is a special gear which only has one tooth and is fixed onto the driver shaft. This then engages the worm wheel which is connected to the driven shaft. If the worm gear has 24 teeth, and the worm has one tooth, then:

Velocity Ratio (VR) = driven gear/driver gear = 24/1 = 24:1

This arrangement would reduce speed 24 times, increase the power by approximately 24 but also has the advantage of changing the direction of rotation through 90°. Worm gears are very compact, with few moving parts, and consequently very reliable.

A *Worm gear*

Bevel gears

If you want to change the direction of rotation through 90°, you could use a bevel gear. The teeth of a bevel gear are cut at an angle of 45° and work in the same way as spur gears.

Rack and pinion gears

These are used to change linear motion into rotary motion, and vice versa. They are found in pillar drills, camera tripods and steering racks.

Helical gears

These gears have their teeth cut at an angle as in a helix, and therefore slightly curved. They are useful for changes in direction and are quieter and smoother than normal spur gears. However they are very difficult to machine, and consequently more expensive.

B *Bevel gear*

Drive shafts

All these systems rely on **drive shafts** to transmit power from one part of the machine to the other. Gears, pulleys and sprockets are mounted on the shafts using a number of methods. For **interference fit**, the shaft and hole are machined to a very high tolerance and rely on friction to stop the gear or pulley spinning. Often, one part is heated or frozen, expanding or contracting the part so the fit is even tighter; however, this is not usually reliable if large forces need to be transmitted. Other methods of mounting gears, pulleys and sprockets onto shafts include grub screws, keys and keyways, and splines.

The components could be glued or welded, but this makes maintenance difficult.

C *Helical gear*

Pulleys and belts

Pulley and belt systems use a belt to transmit motion and power from one place to another, i.e. from the driver shaft to the driven shaft. A wheel has a groove to accommodate a band or belt; both the groove and band have a range of cross-sectional shapes depending on the application and the forces they have to transmit. Speed and power changes are achieved by varying the diameter of the pulleys; for example, a small diameter pulley driving a large diameter pulley reduces speed. If the smaller pulley, the driver, has a diameter of 50mm and the larger pulley, the driven, has a diameter of 400mm:

Velocity Ratio (VR) = 400/50=8:1

In other words, the driven wheel will rotate 8 times slower than the driver.

Chain and sprocket

Chain and sprocket is a very familiar simple machine as it is used on bicycles and motorbikes. As for pulleys and belts, the chain has to be kept in tension. This is achieved on a bicycle by moving the rear wheel forwards and backwards or using a jockey wheel and spring. Chains are much less likely to slip than pulleys.

D *Pulley and belt*

E *Chain and sprocket*

Key terms

Drive shafts: a tube or a rod used for transmitting power.

Interference fit: a method of mounting gears, pulleys and sprockets onto a drive shaft.

Pulley and belt: a wheel with a grooved rim for a rope or band.

Chain and sprocket: a pair of toothed wheels connected by a chain.

∞ links

You may find the following websites helpful:

www.automata.co.uk/pulleys.htm

www.suelebeau.com/gears.htm

en.wikipedia.org/wiki/Driveshaft

Refer back to pages 66–67 for more information about simple and compound gear trains.

Activities

1 Make a list of five products that use a pulley system in their operation.

2 Using a construction kit, build two mechanisms to show how a change in direction of 90° could be achieved.

3 List as many ways as you can of securing a pulley wheel to a shaft. Use diagrams to illustrate your answer.

Summary

A number of specialist gears are available for transmitting drive from one place to another in different ways.

A pulley and belt system, or a chain and sprocket, can also transmit motion and power.

Study tip

You need to know how to connect one drive to another, to achieve changes in direction of drive as well as increases and decreases of speed and power. Very often, that necessitates completion of a diagram, and you should be able to support your choice of mechanism by listing advantages and disadvantages.

If movement is required as the output to a system, motors and solenoids are used. Everyday products would be impossible to produce without the use of these devices.

Motors

Motors come in many shapes and sizes, but the three most common types are the **dc motor,** the **ac motor** and the **stepper motor.** When considering a motor as an output to a system, you have to consider what **torque** the motor needs to produce, its function and its power source. In most cases, in the workshop, either a low voltage dc motor or a stepper motor will be used. These devices need to be provided with their own power supply, and the easiest way to achieve this is by using a **solenoid** within a **relay.**

dc motors

The most popular choice of an electric motor in the school workshop is the dc motor, which can operate on a range of voltages. The miniature motors operate on 1.5V–3V, but a reasonably powerful motor is more likely to operate on 6V–12V. However, these small motors (which are easily obtainable and relatively inexpensive) spin exceedingly fast and consequently have very little torque or power. For school projects it is much more usual to use a dc motor with a compound gear kit or a dc motor with its own built-in gearbox.

ac motors

Because ac motors usually require mains electrical voltage of 220V–240V, they are rarely used in school projects for safety reasons. However, they are used in many domestic appliances that require movement, such as washing machines, vacuum cleaners, refrigerators, drills and microwave ovens.

Stepper motors

Increasingly stepper motors are being used in the school workshop. They have the advantage of moving in steps or increments, very like the second hand of a clock. Typically, they move in steps of 0.9°–1.8° with 400 or 200 steps giving you a full revolution. To control the speed of the device you alter the time interval between each step. They make very precise movements, and also have the ability to move clockwise and anticlockwise. A typical use would be antenna rotators or robot arms.

Solenoids and relays

A solenoid is an electrically energised coil of wire. This generates a magnetic field which in turn attracts a plunger or armature. It is a good example of an energy changer: it changes electrical energy into kinetic energy and produces linear motion.

A *A dc motor*

B *Many common household appliances are powered by ac motors*

Solenoids have many uses, in particular locks, and door or gate releases. However, you are much more likely to come across a solenoid within a relay. A relay is a device that allows you to use a small current to switch on a larger current. Typically, a printed circuit board will only produce on the output side a small amount of current; a relay will then use this current to switch on a motor with a high current demand. Many relays have a clear perspex case and you can see the solenoid move the contact points inside.

∞ links

Take a look at these websites for more information:

www.bcae1.com/relays.htm

www.howstuffworks.com/motor. htm

www.explainthatstuff.com/ electricmotors.html

C *A stepper motor*

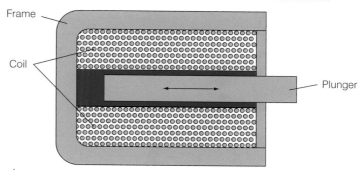

Frame

Coil

Plunger

D *A solenoid*

E *A relay, and how it fits into a circuit*

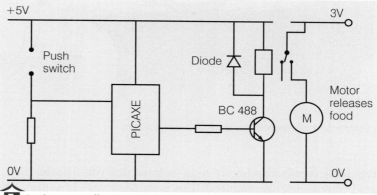

+5V

Push switch

PICAXE

Diode

BC 488

M

3V

Motor releases food

0V

0V

F *A micro controller circuit linked to a motor circuit using a relay*

Activities

1 Use an electronic supply catalogue and list the specifications of three different dc motors.

2 Find a relay with a clear Perspex casing and draw a diagram of the various components you can see inside.

3 Make a list of 10 products that use an ac motor.

Study tip

- It is important that you know how to complete a circuit diagram using a relay.

- In order for a motor to do useful work, you should be able to suggest how simple machines could be connected to the motor spindle.

- Can you suggest a suitable motor for a range of given products?

Summary

There are three main types of motor: dc motors, ac motors and stepper motors.

Stepper motors are often used in the school situation.

A solenoid is a coil of wire that carries an electric current and acts like a magnet. Solenoids are often used in relays in electrical circuits.

5.7 Friction

Depending on the application of the resisting force, **friction** within a simple machine can either be a foe or a friend.

A *Friction can be advantageous – or it can be a real problem!*

If any two services rub together, friction (a force that tries to stop the two parts from moving) is produced. This is an advantage in some machines, like brakes or the belt on a pulley system. In most cases, however, it reduces the efficiency of the machine and increases wear and tear. The first method of reducing friction is to make the surfaces as smooth as possible. The second method is to separate the two surfaces by placing a substance between them, such as oil, graphite, or sometimes even high-pressure air. These materials are called **lubricants**.

Lubrication

The aim of all types of lubrication is to separate two surfaces and stop them from rubbing together; this restricts friction and heat, and thereby reduces wear. There are a whole range of lubricants like oils, greases and graphite, all with different thicknesses or **viscosity** depending on the application. If larger loads or speeds are required bearings are used.

B *Engine oil is an example of a lubricant*

Bearings

The problem of friction can also be overcome by using materials which slide easily over one another. **Bearings** can fulfil this function, and protect the surfaces from wear. These bearings come in a variety of types such as, ball bearings roller bearings and tapered roller bearings as in the illustrations below.

The simplest forms of bearing are bushes or plain bearings, which are made from nylon, white metal, or phosphor bronze. If the loading is light, the bearings could be made from plastic, as in a DVD machine, or metal, as in a car engine. If wear does occur, you may be able to replace the ball bearings rather than a major component.

 Bearings

Activities

1. Find a picture of a bicycle and label all the places where you could find bearings.

2. Given a number of equally shaped and sized blocks, of different materials, design an experiment to test the amount of friction they produce when moving across a surface.

3. Draw and label a diagram to explain what is meant by a ball bearing.

Study tip

With a given application, you should be able to suggest a type of lubricant that would reduce friction or pinpoint an optimum position in a device for a bearing.

links

www.darvill.clara.net/enforcemot/friction.htm is a good physics site with animations illustrating the force of friction.

Information on aspects and videos of friction:
http://videos.howstuffworks.com

Summary

Friction is a force that tries to stop parts from moving when they rub together.

Depending on the application, it can be an advantage or a disadvantage.

Lubricants can be applied to reduce friction.

Bearings that slide easily over each other can reduce friction and wear.

Practice questions

Material and components sample questions

Study tip

Question 1: When asked for a material, do not use wood, metal or plastic; these are categories of materials, not materials. Give a specific material, e.g. pine, steel, polyethylene.

Make sure that you can recognise, name and draw the circuit symbols of electronic components. You also need to understand logic gates.

1   This question is about timer circuits.

The circuit below shows a timer IC used as part of the production line system to act as a timing circuit. The user presses the switch and the buzzer sounds for the timing period.

A

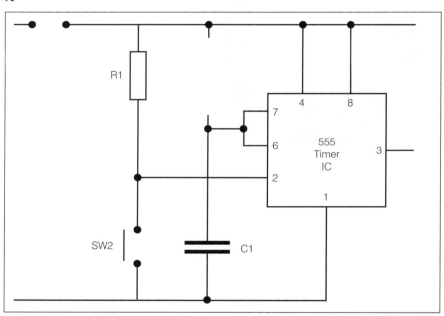

(a)   Add the following circuit symbols to the circuit in Diagram **A**:
  • the 0V and 9V terminations   *(2 marks)*
  • the variable resistor used to calibrate the timing   *(2 marks)*
  • a buzzer to the circuit that will sound during the timed period.   *(2 marks)*

Study tip

Ensure that you do exactly as the question asks; marks will only be awarded for what is asked for.

Take care when drawing electronic circuits and symbols. If there is an error you will not be given the mark.

B

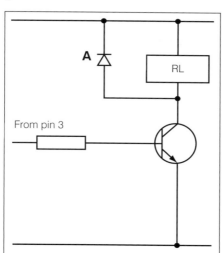

C

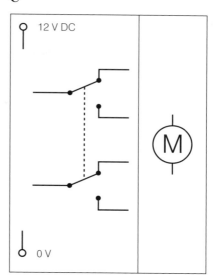

The sound was found to be inadequate, therefore a higher voltage mechanical buzzer was used to improve the circuit and other components were added.

(b) (i) Name component A in Diagram **B**. *(1 mark)*

(ii) Explain why component A is required. *(2 marks)*

2 A double pole double throw (DPDT) switch can be used to control the forward and reverse direction of a motor.

Complete Diagram **C** above to show a reversing circuit. A mark will be awarded for each correct link drawn:

- power supply connected to switch *(2 marks)*

- switch connected to the motor. *(4 marks)*

3 (a) (i) Complete Table **D** by naming the logic gate or drawing the symbol for a NOT or OR gate. *(4 marks)*

(ii) Complete the truth tables for the three logic gates. *(6 marks)*

D

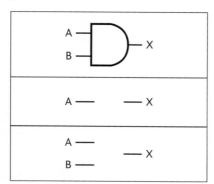

What will you study in this section?

After completing the Design and market influences unit (Chapters 6 and 7) you should have a good understanding of:

product analysis

social, cultural, moral and environmental issues.

Design and market influences

■ Introduction

Design and Technology involves the understanding of design and market influences. You will learn how to analyse systems control products and processes. You should consider how design and technology affects the manufacturer, user and environment, and understand the importance of health and safety.

■ Product analysis

You will need a working knowledge of product analysis. You will learn how to check your design proposals against design criteria and understand the factors which influence design. You will also cover quality assurance through testing procedures, and learn how to check the quality of your work at critical stages of development. It is important that you modify your product during your project to improve its performance.

As your work progresses, you will need to model and prototype it, as well as record your work and planning for continuous evaluation and testing. You need to identify faults during construction and on your complete system. All good designs have come through a series of versions before the final product is released; your project should show this.

A *James Dyson made over 5,000 models and prototypes while developing his bagless vacuum cleaner*

Evaluation of the quality of your system compared with the specification will also be covered. You will learn how to devise an end-user evaluation of your prototype system, which may include a test to check the quality of the system against the specification and ensure its suitability for the intended users.

■ Social, cultural, moral and environmental issues

You will learn how to recognise the effects of social and cultural influences on system design and the impact design may have on society and the environment. This unit covers the effect control systems have had on various occupations (e.g. bomb disposal and car production) and the wider uses of automation.

You need to understand that all systems require maintenance, as well as how to design to meet the maintenance needs of a system. It is important to protect designs and we will look at industry standards for licensing.

Chapter 7, Social, cultural, moral and environmental issues, considers the effects of environmental pollution and waste. You will learn how sustainability can be designed into products at the manufacturing stage, and the difference between renewable and non-renewable energy sources.

Understanding health and safety issues is important, as is learning how to identify hazards and risks when designing and manufacturing products. This applies both to hazards in the workshop and hazards in the real world, and should enable you to make products and systems safer.

B *Good design incorporates easy maintenance as well as visual flair*

C *Wind farms are a sustainable form of energy generation*

6.1 Product analysis

Analysing systems

All systems can be described in terms of their inputs, processes and outputs by using block diagrams. To help us better understand systems we sometimes break them down into smaller parts we call *sub-systems*. In fact, most systems are made up of a number of smaller sub-systems that are connected together to form the main system. Systems can be solely electronic, mechanical or pneumatic, or link together all of these, or combinations of these technologies. To analyse a system or a control product we would look at it firstly as a simple input, process, output block diagram, and then look at any sub-systems within in the same way. To describe its *operation* we might use a **flowchart**.

Types of system

Systems can generally be divided into three categories: manual, semi-automatic and automatic.

Manual systems allow users to make decisions for themselves; an example is the speed control on an electric drill where pressure from the user's finger changes the motor speed.

Semi-automatic systems allow for the majority of the system operation to be carried out automatically but the user has to begin the process; an example is a lift system in an office block where the user has to press a call button to begin the automatic processes.

Automatic systems once started allow the entire process to run continually without any intervention from the user; an example is a domestic central heating system that once it has been turned on will maintain temperatures at predetermined times automatically.

Analysing a central heating system

To begin we could break the system down into a simple block diagram of input, process and output.

∞ links

See pages 26–27 for more information on electronic building blocks.

See pages 37–39 for more about using flow diagrams in systems and control.

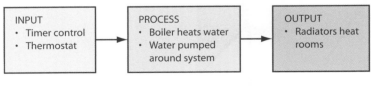

INPUT	PROCESS	OUTPUT
• Timer control • Thermostat	• Boiler heats water • Water pumped around system	• Radiators heat rooms

A *System analysis is best analysed by using flow charts*

B *Timer control*

We could then analyse the sub-systems.

Timer control

This gives control over the times the heating system switches on. Modern controllers also have temperature settings that can be programmed in for different times of the day.

Thermostats

Thermostatic radiator valves can be added to the system to control the temperature in different rooms.

Boiler

This heats the water and has its own systems to ensure the water is heated to the correct preset temperature.

Pump and radiators

This moves the water through the radiators and brings it back to the boiler for re-heating.

We could analyse the *operation* of the system by making a flowchart.

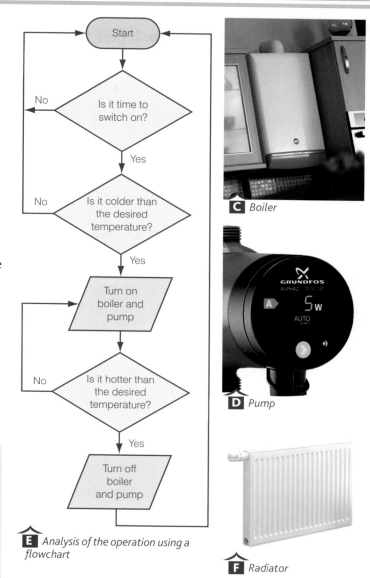

E *Analysis of the operation using a flowchart*

C *Boiler*

D *Pump*

F *Radiator*

Activities

1   Analyse the following products/systems in terms of input, process and output and list any sub-systems and explain what they do:

a   automatic sliding door

b   wind turbine

c   car park exit barrier

d   washing machine

e   lift system.

2   Draw a flowchart for each of the above to describe the operation of each system and state whether the system is manual, semi-automatic or automatic.

Summary

All systems can be broken down and described in terms of input–process–output.

Most systems can be further broken down into sub-systems.

Systems can be described as being manual, semi-automatic or automatic.

System operation is best analysed by using flowcharts.

Study tip

When you analyse a product or system as part of your research, always break it down into input–process–output.

Evaluation techniques and quality control through testing

Evaluating

Evaluating is important when designing products or systems to ensure that the products or systems are designed to meet the intended criteria, that they function and perform as intended, and are made to an acceptable quality.

Specification

I am going to design and make a shop display that attracts people's attention. The following points should be met to ensure the job is done well.

Aesthetics – How it looks from the outside:

- The outside of display should be clean and tidy so that it is able to retain the attention of the passer-by.
- The mechanism (linear actuator) and lights should work properly and to capture the person's attention.

Components – How they should be attached and what voltage supply is necessary.

- Most of the components are soldered to the board – make sure they are in contact with the copper tracks.
- The components that aren't soldered directly to the board – variable resistor, (6) LEDs, linear actuator should be connected with wires that have been twisted round each other to ensure that they are tidy and to help prevent the wires from snapping.
- I will make sure that I use the correct voltage, i.e. 6 volts for the PIC board, 6 volts for the relays and 12 volts for the linear actuator.

Ergonomics – How efficient my display should be:

- The display should be strong so that when handled it won't break.
- The LDR should be placed where the light is able to get to it so that when the customers pass it they will cast a shadow and activate it.
- The components should be fixed into position so that they don't move around and break.
- My display should be designed to suit the needs of the general public – include phone number of the charity.

A *Design criteria for a charity shop display*

Evaluating against design criteria

The specification for a product or system can help the designer determine the characteristics, measurements, functions, appearance and other aspects of the design. The criteria laid down in the specification can be used throughout the design process to ensure the quality of the product. Design ideas can be evaluated against the criteria described in the specification, commenting on how well they match up and therefore which ones to take forward to develop. When the prototype product or system is made, it too can be evaluated against the criteria described in the specification.

Test procedures

It might be helpful to come up with a series of different tests for the product or system to perform. For example, if you were testing a timed battery-operated food whisk, you might try it out with different mixtures.

It is important that all results are carefully recorded, considered and detailed in the final evaluation. All evaluation and testing helps to make sure that the final product or system produced is of an acceptable quality, and performs as intended.

End user testing

Products or systems are designed with the people who will use them in mind. It is important, therefore, that the views of these end users are found out so that the level of success of the design can be established. One way of doing this might be to ask the end user a series of questions after they have used or operated the product or system. The design criteria listed in the specification should list what the product or system is intended to do; the **closed questions** the end user is asked would help evaluate the level of success of the product or system.

User Questionnaire

	Yes	No
1. Would the Charity Box encourage you to donate money?	☐	☐
2. Did the Charity Box activate automatically when you approached?	☐	☐

3. How attractive do you find the appearance of the charity box?

 1 2 3 4 5
 ☐ ☐ ☐ ☐ ☐

 1. Very attractive ⟶ 5. Not attractive

4. How effective do you find the flashing lights at attracting you?

 1 2 3 4 5
 ☐ ☐ ☐ ☐ ☐

 1. Very attractive ⟶ 5. Not attractive

B *End user questionnaire*

Summary

Evaluating helps ensure the product or system functions and performs as intended.

Evaluating helps ensure that a quality product or system is produced.

Test procedures can be used to help evaluate.

End user questionnaires can help ensure customer satisfaction.

Activity

1   In groups discuss and work out how the success of a new washing machine could be evaluated.

a   Think about what the criteria for a successful washing machine might be and list them.

b   Devise a series of tests you might perform and list them.

c   Produce a questionnaire consisting of closed questions you could ask end users.

Key terms

Closed questions: questions which give a limited choice of answer.

∞ links

Find out more about testing and evaluation in Chapter 12.

C *Electric whisk*

Study tip

■ Ask experts or end users their opinions to test your ideas, proposals and prototypes.

■ Always look back at any specification you have written or been given when evaluating any idea or end product or system.

7 Social, cultural, moral and environmental issues

7.1 Social, cultural, moral and environmental issues

Social influences on system design

Products and systems are all designed to meet the needs of individuals and society as a whole. The Baylis clockwork radio in the case study gives an example of how one inventive person responded to the needs of a very different society from his own.

Since the first clockwork radios were produced, many other green products and systems have been designed and produced. There is now global concern about the environment and how we cater for our energy needs. Consumers are looking at alternative energy supplies to power their products and systems, and designers have to pay increasing attention to the need for **sustainable design**.

Objectives

Recognise the effects of social and cultural influences on system design.

Learn about the impact design may have on society and environment.

Learn about the effect control systems have had on various occupations.

Case study

At home in the UK in the early 1990s, Trevor Baylis watched a television programme on the spread of Aids in Africa. He learned that the only means of communication in many areas was radio, but electricity was either unavailable or too expensive, as were batteries. Education was desperately needed, but the technology to meet this need was not available.

Trevor Baylis started researching and experimenting, and produced a clockwork radio, using a mixture of old and emerging technologies.

The African Liberty Life Group heard about Baylis's design and were able to make the clockwork radio a reality when the South African company Baygen started production in 1994. A smaller, lighter and transparent version was produced for the Western market in 1997.

A *The Baylis clockwork radio*

The impact of design on society and the environment

A hundred or more years ago, working in a factory was not pleasant. The work was hard, repetitive and often dangerous too. Conditions were cramped and crowded and there was no health and safety legislation.

Today, many unpopular, repetitive and monotonous factory jobs have been replaced by automated systems and the use of robots. Whether this has been a benefit to society or not is a point for debate. Many boring jobs have been lost, production has speeded up, prices of goods have generally come down as a consequence of cheaper labour costs and faster production. New jobs have been created in manufacturing and maintaining new automated systems; more energy is required to run the new automated systems.

Activity

1   Read about Trevor Baylis's clockwork radio and explain why the specification for radios to be sold in the West was different to the African version.

Key terms

Sustainable design: designed in an environmentally friendly way and so that it will last.

B *Working conditions in factories were very poor in nineteenth-century Britain*

C *Industrialisation brings advantages, but it also brings pollution*

∞ links

Check out pages 90–91 for more on sustainability and environmental concerns.

Remember

Automated systems and new technologies may take away some jobs but others are created.

D *In modern factories, many boring, dangerous or repetitive tasks can be done by robots*

Activity

2 Working in small groups, list as many systems as you can that have benefited your lives – for example, the central heating system. Discuss the impact these systems have had on society and whether or not they have impacted on the environment.

Study tip

■ Make sure you can explain the effects automation has had on people's lives and jobs.

■ You should be able to give examples of how new technologies can affect society.

Summary

Products and systems are designed to meet the needs of societies.

Automated systems can do repetitive tasks quickly and efficiently instead of humans.

Some jobs disappear and others are created with new automated systems and technologies.

Product maintenance

Recognising maintenance needs

All systems or products require **maintenance** to ensure that they continue to operate efficiently and as designed. If a system is not maintained and it fails, it could be expensive to fix or replace – or it might become very dangerous.

The amount and frequency of maintenance varies depending upon the system, its use and the environment it is operating in. Maintenance requirements are affected by:

- the sensitivity of the product
- the needs of the user
- the environment in which it is operating.

Minimising wear

As a product is used, parts wear out due mainly to friction, which is generated when objects rub together. Several steps can be taken to combat this.

Oiling

Oil is used to **lubricate** the parts of mechanical systems. The oil allows the parts to slide across each other, preventing friction from causing damage.

The oil should be filtered to catch microscopic metal particles that could cause damage. Oil can be added to the moving parts of a product, such as a bike chain. More complex systems, such as a petrol engine, have an oil pump, oilways and a sump to catch the oil. Oil is constantly pumped around the engine to ensure all moving surfaces are lubricated.

Replacement of parts that wear

Some parts are not replaced until they fail completely, like light bulbs. In critical applications, such as a lighthouse or on a rally car, spare bulbs are installed and can be quickly swapped into the system.

Some parts that wear cannot be oiled and have to be replaced when they wear out. Examples include brake pads or clutch pads in cars. These items rely upon friction to slow the vehicle or transmit drive.

Other parts wear out because they are on the outside of the product and used to interact with other materials. Examples on a motor vehicle include car tyres and wiper blades.

Adjustments

Sometimes the system itself can adjust to a worn part's smaller shape. This adjustment can be automatic or controlled by the operator.

An example of an automatic adjustment is on a disc-brake system on a mountain bike. As the brake pad wears it becomes thinner, the brake calliper moves out to take up this gap and the brake continues to function. The calliper continues to push the pad onto the brake disc

Key terms

Maintenance: the work of keeping a product in good condition.

Lubricate: to make slippery.

Service: a routine inspection and the carrying out of maintenance activities.

Service interval: the time between services.

A *A mine-cutting tool needs constant maintenance*

B *A car tyre will wear and need replacing*

⊙⊙ links

Refer back to pages 72–73 to remind yourself how to combat friction.

even after the pad has worn away. Modern cars have sensors in the pads to warn of this happening.

Inbuilt maintenance ability

Designers should ensure that their product or system can be easily maintained. There should be adequate access to parts that need to be checked or replaced (service hatches can be seen on lampposts, printers etc). Parts that fail regularly (blades on a power tool, light bulbs) need to be changed without the use of tools or operator training.

The Dualit toaster has been designed so that a trained technician can dismantle it and replace the elements that fail over time. Another example of a product that is relatively easy to repair is the Land Rover. Over half the Land Rovers produced are still in use: few other car manufacturers can make that claim.

Service intervals, system self-monitoring

In a simple system or product, maintenance may involve no more than replacing a failed component. A more complex product would have specific maintenance requirements at each **service** and specific **service intervals**. These are usually time-based or (as with cars) distance-based. Some service intervals or statutory tests are defined by law: all cars over three years old must take a regular MOT test to ensure that key parts are unlikely to fail.

C *A disc-brake system*

D *The Dualit toaster is a good example of a product that can be easily maintained*

> **Remember**
>
> Products should not be dismantled by unqualified persons. Do not open appliances!

Activities

1 Discuss the advantage to the manufacturer of making a product easy to maintain or throw away.

2 List products found in the home that can be maintained and those that have to be replaced when they fail.

3 Discuss whether the government should encourage users to maintain rather than throw away.

4 Identify environmental issues of the 'throw away' society.

Summary

The majority of systems require maintenance.

Lack of maintenance can be costly and dangerous.

Maintenance includes oiling, replacement of parts that wear, adjustments.

Systems should be designed to allow them to be maintained.

Systems have recommended service intervals, i.e. the time between services.

Complex systems can offer self-monitoring where they define the service intervals.

links

To discover more about the issues of throwing items away visit:

www.greenpeace.org.uk

www.environment-agency.gov.uk

> **Study tip**
>
> ■ Ensure that you can recognise the maintenance needs of a system.
>
> ■ You need to be able to explain oiling, replacement of parts that wear and adjustments.
>
> ■ You need to understand how to incorporate the ability to maintain a system in your design.
>
> ■ Make sure that you can discuss issues like the 'throw away' society.

The advantages of protection

There are many simple and cost-effective ways for inventors to protect their ideas. These are known collectively as intellectual property and are monitored and protected in the UK by the Intellectual Property Office. If an idea is not protected, there is little to stop another person or company copying the inventor's original idea and making money from it.

The process of protecting an idea can take many months.

Forms of protection granted by the Intellectual Property Office

The Intellectual Property Office grants protection but it does not help fight any breach of this protection. That has to be done through the courts. Ideas and products are legally protected in the following ways:

Patents

A **patent** protects new inventions. It can cover what the device is, how the device works, what it is made from and how it is manufactured. It gives the owner the right to prevent others from making, using, importing or selling the invention without permission. A patent can be for a whole product or for just part of a product. Dyson Ltd has around 1,200 patents registered for different vacuum cleaners and their parts.

The invention must:

- be new
- have an inventive step that is not obvious to someone with knowledge and experience in the subject
- be capable of being made or used in some kind of industry.

A patent lasts for five years but can be renewed for up to 20 years' protection.

Trademarks

Trademarks are symbols or logos that distinguish products. A trademark must be distinctive for the goods and services you provide.

Designs

Designs are not inventions but a new unique style of existing products, e.g. a design of training shoe or handbag.

A registered design is a legal right which protects the overall visual appearance of a product in the geographical area you register it. The visual features that form the design include the lines, contours, colours, shape, texture, materials and the ornamentation of the product which give it a unique appearance.

A The Dyson DC25

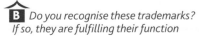

B Do you recognise these trademarks? If so, they are fulfilling their function

To be eligible for registration, a design must:

- be new
- have individual character.

Copyright

Copyright is automatic and applies when the work is written or recorded in some way.

Copyright can protect:

- literary works, including novels, instruction manuals, computer programs, song lyrics, newspaper articles and some types of database
- dramatic works, including dance or mime
- musical works
- artistic works, including paintings, engravings, photographs, sculptures, collages, architecture, technical drawings, diagrams, maps and logos
- layouts or typographical arrangements used to publish a work, for a book for instance
- recordings of a work, including sound and film
- broadcasts of a work.

Breaking protection laws

If a person owns an intellectual property right – a copyright, design, patent or trademark – no one else can manufacture it without their prior permission. Unauthorised use of someone's intellectual property is a crime and may lead to prosecution.

Counterfeiting relates to wilful trademark infringement. Examples of counterfeiting include fake clothing, prescription drugs, footwear, etc.

Piracy relates to wilful copyright infringement – for example, by copying films, music, software, computer games, and so on.

If a person feels that their intellectual property is being infringed, it is up to them to sue the offender.

Licensing

Licensing allows inventors and manufacturers to agree terms when working together. The inventor is paid an agreed amount for his idea and the manufacturer is able to produce the products without prosecution. This allows the inventor to have someone else manufacture the product, and the manufacturer can make money by legally producing someone else's product.

The Intellectual Property Office offer advice about arranging licences. An intellectual property licence gives permission to do something that, without the licence, would be an infringement of intellectual property.

Activities

1 Research the cost and process to patent an idea in the UK.

2 Investigate successful prosecutions of counterfeiting or piracy cases.

3 Search for patents online for a specific area, such as folding bicycles.

⬤⬤ links

The list on the left is taken from the Intellectual Property Office website: **www.ipo.gov.uk**

⬤⬤ links

To discover more about protecting your ideas, visit:

www.ipo.gov.uk

www.crackingideas.com/site

www.piracyisacrime.com

Study tip

- Ensure that you understand the advantages of protection.
- Know the difference between patent, trademarks, design and copyright.
- Be able to give the advantages of licensing to inventors and manufacturers.

Summary

If intellectual property is not protected anyone can copy it without being charged.

The Intellectual Property Office grants protection in the UK.

Patents, trademarks, design and copyright are forms of intellectual property protection.

Licensing allows inventors and manufacturers to agree terms when working together.

Environmental pollution

Our planet has suffered from **environmental pollution** since the industrial revolution. The creation of factories and the consumption of huge quantities of coal and oil created air pollution never seen before. From the late nineteenth century, laws have been passed in the West to attempt to reduce pollution. The developing world has yet to adopt such stringent regulations.

Air pollution involves the release of chemicals, dust or smoke into the atmosphere (e.g. from chimneys, chemical plants).

Light pollution involves excessive light at night (e.g. floodlighting on an industrial complex).

Noise pollution involves excessive noise (e.g. aircraft noise, noise from heavy industry).

Radioactive contamination is nuclear contamination (e.g. from nuclear waste or weapons research).

Soil contamination involves chemicals being released into the soil (e.g. leakage from underground tanks).

Thermal pollution is the raising of the ambient air or water temperature (e.g. the cooling of a power station can warm the air or sea).

Visual pollution spoils the original views (e.g. by pylons, rubbish tips).

Water pollution is caused by waste products and chemicals being released into rivers and the sea.

B *Water pollution*

C *Visual pollution*

Environmental costs

Rainforests

Gold, copper, diamonds and other precious metals and gemstones are important natural resources that are found in rainforests around the world. Extracting them is frequently destructive, damaging the rainforest ecosystem and causing problems for people living nearby and downstream.

Objectives

Learn about the effects of environmental pollution.

Understand the uses and wider effects of automation.

Key terms

Environmental: belonging to the natural world: land, sea, air, plants and animals.

Pollution: the introduction of contaminants into an environment.

Production line: mechanical track in a factory upon which products are assembled.

Robot: a device that is operated by remote control or computer program.

A *Air pollution*

∞links

To discover more about environmental issues visit:

www.draxgroup.plc.uk/explore_drax

www.greenpeace.org.uk

www.cat.org.uk

www.sas.org.uk

campaigns.direct.gov.uk/actonco2/home.html

Climate change

The world is warming up due the burning of fossil fuels. Global weather systems are affected – as the polar ice caps melt, sea levels rise causing flooding.

Toxins

The majority of our domestic products and consumables contain toxins. Toxins in items disposed in landfill sites can leach into the soil and pollute the water table.

Oceans

The oceans are becoming polluted, threatening fish and smaller organisms. When an oil tanker runs aground and breaks up it makes headline news, but far more damage is being done invisibly on a daily basis.

D *A car production line*

Nuclear

Nuclear power stations create a small amount of waste that is highly dangerous and will stay that way for many thousands of years. There have been nuclear accidents at Three Mile Island in the US and Chernobyl in the Soviet Union.

Automated production lines

In a car **production line**, the car grows from the bare steel panels that are welded together to a finished vehicle that is driven off the production line a few hours later. In some industries the production line never stops and the workers man the line in shifts. If there is demand for the product, a production line is the cheapest form of manufacture. Disadvantages include boredom amongst workers.

Common applications of robotics

Robots are used in industrial processes to replace workers. This is either to protect the worker or because the robot is less prone to error. Some mobile robots are controlled directly by an operator, who can stay away from danger. These are useful for bomb disposal units, as seen in Photo **B**, nuclear power plants and deep-sea investigation. Other robots are controlled by computer to repeat the same process quickly and accurately. They are useful for modern production lines.

Summary

The world has never been so polluted as now.

There are many types of environmental pollution.

Industry can take steps to minimise its environmental pollution.

Automated production lines can be the cheapest way to produce products.

Robots can be used to safeguard the workers and complete repetitive complex tasks.

E *Robots can be used for risk-free bomb disposal*

Activities

1. Discuss in groups the damage that manufacturing in the West has done to the environment.

2. Investigate serious natural disasters. Research 'Amoco Cadiz', 'Chernobyl', 'Three Mile Island' in an internet search engine.

3. Investigate ways that your home or school could reduce its carbon footprint.

4. Discuss the potential winners and losers of the introduction of robotic production lines.

Sustainable design, sometimes called **green design**, is a method of manufacturing systems and products which comply with economic and ecological sustainability, particularly machines and systems that are energy-efficient and are environmentally friendly.

Sustainable manufacture

When designing systems and products, the issues to consider are:

- materials: they should be non-toxic, recyclable, sustainably produced and use less energy to be manufactured
- low carbon footprint, minimising CO_2, both in the manufacture and delivery, perhaps produced locally
- design for re-use, made to last longer and be replaced less frequently
- design to use less energy, perhaps using renewable energy sources like solar power.

Greener mobile phones

Consider the sustainable manufacture of mobile phones. PCBs use many hazardous substances, including lead, cadmium, mercury, bromide compounds and arsenic and the components should be carefully examined in regard to **recycling** and reuse. PCBs come directly under the waste electrical and electronic equipment directive (WEEE 2003) and under Landfill Regulations (2002) that set targets for IT equipment such as LCD displays, PCBs, batteries and flame-retardant plastics. Casings, rather than being made from oil-based plastics, could be made from plant-based plastic and consequently biodegradable. You could tap into solar power and make them more energy-efficient. Finally extend their expected life – rather than renewing every 18 months, try to renew them after three years, thereby reducing their overall environmental impact.

Recycling

Recycling is the process of using used materials to make new products in order to prevent waste, reducing the need for using fresh raw materials, and thereby reducing the energy usage needed in their manufacture. This in turn reduces water pollution, air pollution and the need for landfill.

Some materials are relatively easy to recycle, like aluminium or glass. Other things like PCBs are much harder and need to be broken down to their component form. Many materials can be changed directly; for example, drink cans can be directly recycled into a new drink can. Other materials are harder to salvage, for example magazine paper, which is usually turned into packaging or cardboard. Sometimes recycling occurs because the materials are intrinsically valuable, like lead from car batteries.

Objectives

Understand the importance of sustainability.

Learn how sustainability can be designed in the product at the manufacturing stage.

Key terms

Recycling: using previously-used materials to make fresh products.

Obsolescence: the period of time after which a product ceases to function.

Product life cycle: the various stages in a product's lifespan.

A *How do you make mobile phones sustainable?*

B *Recycling box*

Obsolescence

Some products become non-functional after a period of time or use. This is called **obsolescence**. For some manufacturers, that is an advantage; a light bulb might have a limited lifespan, and consequently the consumer has to buy more. However, the use of planned obsolescence by producers is very difficult to assess; you could make a very long-lasting light bulb that could last generations, but it would be quite expensive, and the manufacturer would soon be out of business.

Product life cycle

The various stages with a manufactured item's lifespan are referred to as the **product life cycle**. This covers introduction into the market, expanding market growth, maturity (when you have a number of competitors) and decline. The introduction of electronic notepads is a good example: word processors, the growth of PCs and laptops and, finally, the decline of typewriters.

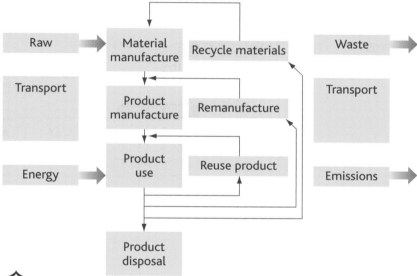

C *Product life cycle flowchart*

Activities

1. Explain the meaning of the following terms; carbon footprint, fair trade, product miles. Pick three everyday products, either from home or the school workshop and explain how they might be recycled.

2. Identify five products which have, in your opinion, planned obsolescence.

Study tip

Consider how manufacturers improve sustainability of one of their products. Review the issues you considered when producing your project coursework. Could you readily identify components in a product that could easily be recycled?

Summary

You should have a basic knowledge of the principles that govern sustainability, recycling and obsolescence.

You should be able to identify opportunities in a product to lessen their environmental impact on society.

7.6 Energy sources

Energy is often defined as the ability to do work or cause change. Energy is all around you and exists in different forms. It can neither be created nor destroyed, only changed from one form to another.

Objectives

Understand the difference between renewable and non-renewable energy sources.

Be able to list all the different forms of energy and how they are changed from one form to another.

Forms of energy

Energy can take any of the forms indicated in the table below.

A *Different forms of energy*

Kinetic	Moving energy	e.g. a train moving along a track
Potential and chemical	Stored energy	e.g. a coiled spring, energy stored in a fossil fuel
Thermal	Heat energy	e.g. energy given out by a warm car engine
Light	Solar	e.g. energy given out by the sun
Sound	Vibration	e.g. noise from a loudspeaker
Nuclear	Atomic energy	e.g. the release of energy as result of nuclear fission
Electrical and electromagnetic		e.g. energy given out by a battery

Energy transformations

If energy is being used then work is being done. This usually involves changing energy from one form into another (known as energy transformation). There are countless examples, for example:

- A light bulb converts electrical energy into heat and light.
- A car turns chemical energy, i.e. petrol, into kinetic energy (as well as heat, sound, electrical and light energy).
- A dynamo converts kinetic energy into electrical energy, and if it is then connected to a lamp it will be converted from electrical energy into light energy.

Key terms

Renewable energy: energy generated from natural resources which are naturally replaced.

Non-renewable energy: energy resources that are not replaced immediately.

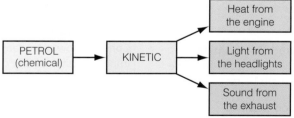

B *Energy transformation*

Energy sources

We categorise energy sources into finite and non-finite sources, i.e. **renewable** and **non-renewable**. A non-renewable source will run out because it is made over such a long period of time. The best examples are fossil fuels, oil, coal and gas. Renewable energy sources include wind, solar, geothermal and tidal energy. These are constantly being replaced.

Energy sources are controversial and the subject of an ongoing debate. Difficult issues include the toxic waste generated as a side product of nuclear energy, the amount of energy required to make a wind generator and how long it takes to get a return, the impact of a tidal barrage on the ecological environment.

The efficiency of a system can be measured as a product of energy:

efficiency = energy out/energy in × 100%

C *Wind generators*

D *Bubbling lava pool*

E *Oil rig*

F *A hydroelectricity power station*

Study tip

- Can you give examples of energy transformations?
- Can you suggest alternative energy sources to fossil fuels and list their advantages and disadvantages?
- Can you identify the energy changes in a gas-powered power station?

Currently, we are constantly trying to make systems and machines more efficient. Efficiency measures include house insulation, electric cars, solar panels, etc. It is critical that we examine all the issues involved. We have to take a holistic view of energy systems and manufacture, for it is critical we consider what mistakes we may be making for future generations.

Summary

Energy exists in many different forms. It cannot be created or destroyed, but it can change from one form to another (energy transformation).

Energy sources are infinite or renewable (e.g. from the sun, wind, tides) or finite and non-renewable (fossil fuels – gas, oil). Sooner or later non-renewable sources will run out.

∞links

Ideas and information in saving energy, especially in the home: **www.energysavingtrust.org.uk**

Energy issues and the future: **www.sciencemuseum.org.uk/energy**

The Centre for Alternative Energy, North Wales. All aspects of renewable energy sources can be found at: **www.cat.org.uk**

Health and safety issues

Health and Safety are an essential aspect of system design. Health and safety issues fall into three categories: safety in systems manufacture, safety in systems operation or function, and safe disposal.

Making the system

Workshops can be very dangerous places. Because of this, we have to have a number of rules to keep the environment safe. Most rules can be put down to common sense, but it is useful to list the most obvious:

- Always wear safety glasses when using machinery.
- Anything loose, such as clothing or hair, needs to be tied back.
- Never dash about the workshop.
- Do not operate machinery unless you have permission.
- If you have any concerns about health and safety, ASK your teacher.
- It is the responsibility of the designer and maker to make sure that, during the manufacture of a product or system, no harm can come from the materials and processes being used.

Animal hazard

Sharp instrument hazard

Heat hazard

Glassware hazard

Chemical hazard

Electrical hazard

Eye and face hazard

Fire hazard

Biohazard

Laser radiation hazard

Radioactive hazard

Explosive hazard

A *Warning symbols are displayed for a reason. Make sure you know what they mean and take notice if you see them!*

Using the product

As well as considering health and safety in manufacture, you must consider these issues when the product is being used. Are there any

B *Always wear safety glasses when using machinery*

C *If a part of a machine is guarded, there will be a good reason*

D *The British Standards Institute's Kitemark shows that a product meets safety standards*

sharp edges? Are the materials non-toxic and suitable for use? Are all the components sufficiently guarded, the wires insulated, moving parts covered, and is the structure strong enough for the purpose intended?

Disposal

In the recycling of any system or product, health and safety issues have to be considered, e.g. the release of dangerous gases or harmful processes. Can the product be dismantled into its component form easily and safely, and can it be disposed of without causing harm to anyone and the environment?

E *How many products can be disposed of without causing any harm to the environment?*

Summary

Health and safety issues fall into three categories: safety in systems manufacture; safety in systems operation or function; and safe disposal.

There will be safety rules and procedures in the workshop – observe them.

If you are designing a product, you will need to think about how it can be used safely, and how it will be disposed of eventually.

Activities

1 Pick three different machines commonly found in the school workshop and write out a list of safety precautions. Design **either**:

■ a poster encouraging pupils to work safely in the school workshop

■ a poster detailing what aspects must be considered when designing a system or product. Find out and draw as many safety symbols as you can.

2 What is a Kitemark?

⚬⚬links

The Royal Society for the prevention of accidents:
www.rospa.com

Health and safety using various tools and processes:
www.technologystudent.com/health1/ed1.htm

Home of the BSI Kitemark:
www.bsi-global.com/en

Practice questions

Design and market influences sample questions

This question is about design processes.

The owner of the theme park has asked you to design an automated mascot to welcome visitors to the theme park (automate means to control, or operate automatically).

1   (a)   This part of the question is about analysis.

List two things that you should think about when designing the automated mascot and give a reason why this is important. An example has been given.   *(4 marks)*

Example: The likely cost of the whole project so that we can sell it at a reasonable price.

(b)   This part of the question is about research.

The layout of a research plan for the automated mascot is shown in Diagram **A**.

Complete the diagram below by adding suitable research sources and stating the information that you would hope to find. The materials section has been completed for you.   *(7 marks)*

A

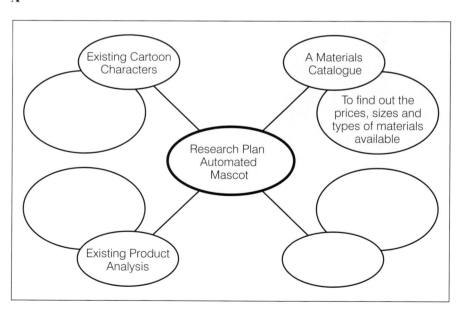

- Existing Cartoon Characters
- A Materials Catalogue
  - To find out the prices, sizes and types of materials available
- Research Plan Automated Mascot
- Existing Product Analysis

(c) Describe how the information from analysis and research may affect the final design. *(2 marks)*

(d) Design specification: give and explain four design requirements for your automated mascot. *(4 × 2 marks)*

Example:
- Requirement: the mascot must welcome visitors.
- Explanation: the mascot must move automatically when the visitors approach it.

2 This question is about design ideas.

Study the information given above and in your design specification.

Use this information to sketch two different design ideas for an automated mascot.

You must show:
- paths of movement of any parts that move
- what powers any moving parts, e.g. mechanism and motor(s) or pneumatic valve(s) and cylinder(s)
- notes to explain your ideas
- evaluation of each idea.

Marks will be awarded for:
- two different ideas   *(2 × 6 marks)*
- quality of sketches/colour   *(5 marks)*
- quality of notes   *(3 marks)*
- quality of evaluations.   *(2 × 3 marks)*

(26 marks in total)

3 This question is about design issues for railway systems.

(a) (i) What advantages would an automatic door system on a train give the passengers? *(1 mark)*

    (ii) Give two reasons for your answer. *(2 marks)*

(b) (i) What advantage would an automatic train door system give the train company? *(1 mark)*

    (ii) Give two reasons for your answer. *(2 marks)*

(c) Suggest how passengers could escape from the train if the automatic door system fails. *(3 marks)*

(d) State two maintenance requirements of the automatic train door system. *(2 marks)*

(e) Give two reasons why electric trams are more environmentally friendly than most cars. *(2 marks)*

(f) A local council wants to encourage car drivers to use trains or trams. Suggest two ways that they could do this. *(2 marks)*

Industrial processes and manufacture

■ Introduction

Design and Technology involves the understanding of industrial processes and manufacture. You need to understand common processes that are used in industry even if you do not have access to them for your project. Knowledge can be gained from the internet or from documentaries.

You should be aware of, and use as appropriate, manufacturing processes and techniques, including computer-aided design (CAD) and computer-aided manufacture (CAM), for systems and control products.

■ Industrial processes and manufacture

It is important to understand that clients, designers, manufacturers and users all have a role in the development of products for industrial manufacture. You will learn how new designs are developed using consumer surveys, questionnaires and technological advances. The chapter also covers different manufacturing processes and techniques, helping you to understand which type of production is appropriate for a given product or system.

■ Systems and control

You need to understand and implement the concept on input–process–output and know how to interpret flowchart diagrams that describe a sequence of events. You will also cover sub-routines and feedback loops.

■ Industrial design and market awareness

You need to be aware that clients, designers, manufacturers and users all have a role in the development of products for industrial manufacture. You may be asked to explain the role of market research and technological advances in the development of a specific design, such as an iPhone.

A *Robots have taken their place on the production line*

B *Market research*

Information and communication technology

You need to understand where ICT is used in the commercial manufacturing processes. You should be able to use ICT software packages as appropriate to research, collect, sort and present research information in your portfolio. Use CAD to generate and develop your design proposals and final specifications, for example to produce PCB mask on a computer. Wherever possible, use CAM to enhance accuracy and quality when producing components and products.

8.1 Industrial processes and manufacture

Your coursework project is an example of a **prototype** – that is to say, it is a 'one-off'. As part of your course, it is important that you are aware of how your product or system might be commercially produced, and the various methods of production that could be used to produce it in quantity. In addition to prototypes, these include **batch production**, **mass production** and **continuous flow production**.

Prototypes

These are individual pieces made one at a time. Prototypes are very often made to test whether a product is viable, whether it works and has any market potential. However, it might be that a product is made for one particular purpose, a handmade clock for instance. Prototypes are very time-consuming and therefore expensive and may be considered specialised items.

Mass production

This is when a product is made in large quantities and in stages along a line, a method of production pioneered by Henry Ford in the 1920s. The base of the product is built first and then components are added by specialist workers as it progresses along an assembly line, finally ending up with a finished article. This could take hours, days or months. This is a relatively inexpensive production process.

A Henry Ford is credited with developing the production line for mass production of his cars. This form of production is still in use in the automobile industry today

Key terms

Prototype: one single item, sometimes built to test a product.

Batch production: small quantities of a similar product made at the same time.

Mass production: a production process in which large numbers of the product are made in stages and finally completed as a whole.

Continuous flow production: the product is made continuously over a period of time, which might be years.

Study tip

- Given a picture of a product could you identify what method of production was used to make it?
- Remember to consider how products or systems might be manufactured commercially.

Batch production

In batch production, similar products are made in small quantites, using the same methods and materials. Jewellery would be a good example.

B *Jewellery is often made in small batches*

Continuous flow production

This form of production involves products being made continuously, with the manufacturing machinery working 24 hours a day, seven days a week. Products made in this way are usually simple in design and straightforward to manufacture, such as nails and bolts. They include commodities such as chemical and steel.

C *Nails are an example of products which can be made by continuous flow production*

Summary

The making of products can be organised in various ways.

A prototype is a 'one-off' item made on an individual basis often for testing purposes.

Some products are made in small batches, with each batch using the same tools and materials.

Larger numbers of products can be mass-produced on a production line, or produced on a continuous basis where the machinery never stops.

Activities

1   List three different products that would be suitable for batch, mass and continuous flow production.

2   Compile a list of the stages involved in the mass production of a bicycle, starting with the painted frame.

∞ links

BBC website detailing industrial processes: **www.bbc.co.uk/schools/ gcsebitesize/business/production/ methodsofproductionrev1.shtml**

Excellent site giving all the production methods with useful animations: **www. notesandsketches.co.uk/Industrial- production.htm**

Shows a production line of a bicycle with video: **videos.howstuffworks. com/howstuffworks/4168- assembly-line-cannondale-bicycles- video.htm**

Flowchart diagrams

Flowcharts allow designers to describe the operation of a system or systems in simple terms. They help designers think through the sequence of events in the system and are another way of modelling ideas. They show how the control system will operate without showing how the operations are actually going to be carried out. There is a set of graphical symbols that have become the accepted convention when drawing flowcharts.

Symbol	Description
START	Used at the beginning of the control process
END	Used at the end of the control process or to indicate the termination of system operation
INPUT/OUTPUT	Shows inputs to the system and outputs from the system
PROCESS	Used to describe what the control system is doing
DECISION	Used to show a decision has to be made in the system at this point
SUB-ROUTINE	Used to show a sub-routine. A sub-routine is a sequence of operations within the larger operation
RETURN	Used to return to a pre-determined location in the operation

A *Flowchart symbols*

Interpreting flowcharts

The flowchart right describes the operations carried out in a drinks vending machine offering tea and coffee.

The flowchart for the main operating system describes clearly how the control system will operate. When a **sub-routine** is reached, this can be described in a separate chart and those operations are carried out before returning to the main flowchart. This gives a good graphical representation of the control system and can also be used in other applications such as computer software for programming PIC microcontrollers, and in planning out making your project.

Objectives

Understand how to interpret flowchart diagrams describing a sequence of events.

Learn how to draw flowchart diagrams describing systems.

Key terms

Sub-routine: the operation of a sub-system within a system.

Activities

1 Draw a flowchart to describe the operation of a security lighting system.

2 In groups, discuss the stages you went through when making your last practical project in D&T in school and draw a flowchart to describe the process. You will need to use flowchart symbols in the appropriate shapes and arrows to link the system. Include any sub-routines in the flowchart.

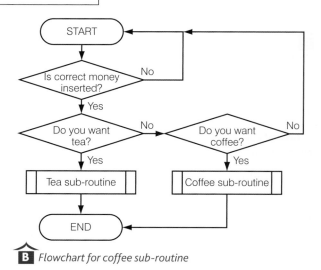

B *Flowchart for coffee sub-routine*

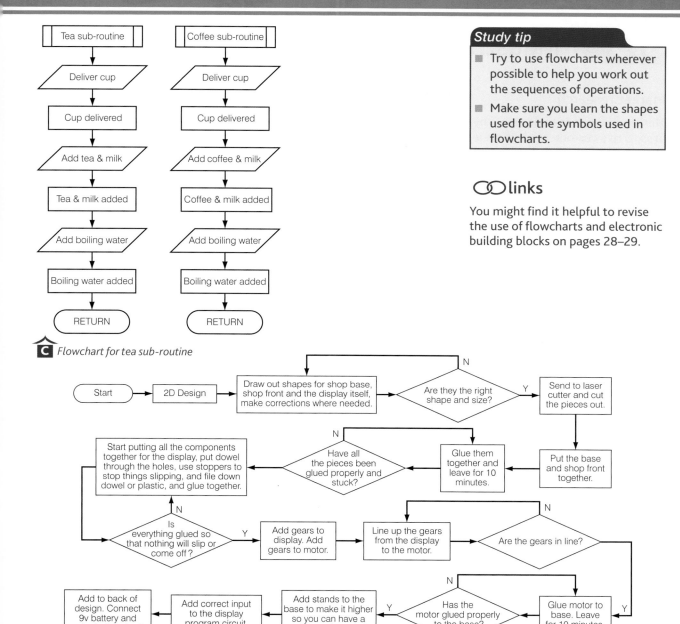

C Flowchart for tea sub-routine

∞links

You might find it helpful to revise the use of flowcharts and electronic building blocks on pages 28–29.

D How one student has used a flowchart to aid production of her project

Summary

Flowcharts are used to describe the operation of a system.

A sub-routine can be used to describe the operation of a sub-system.

Flowcharts can be used in other applications, for example programming PIC microcontrollers and in project planning.

Market research

Market research is undertaken by companies to find out what their prospective customers want or need from their future products or services. A **consumer survey** is undertaken to gain understanding of the needs of existing or future users. It also involves investigating what the company's competitors are currently doing or offering. It is important to conduct accurate market research as this will direct the company's decisions as to which products and **markets** to invest in. Accurate market research ensures that the company's next product or service will be well received by potential customers, and the money that has been invested will be returned.

Methods of conducting a survey

Market research can be conducted in various ways. A popular method is to pay people to stop members of the public in the street and ask them to complete a questionnaire. An advantage of this method is that they can choose people to stop who fit the user profile of the product, e.g. teenage girls; a disadvantage is that it is expensive. Other methods include telephone or online questionnaires, customer workshops or monitoring buying trends with store cards for example. One of the major problems with market research is its reliability; people being asked may be lying or change their mind afterwards.

Assuming that there is confidence in the market research undertaken, this can steer the company in its decisions regarding the **product development**. If the market research suggests that the public would buy the future product or service, this would encourage the company to invest in its development. Similarly, if there were little interest in the product, the company could avoid losing money by not investing in further development of the product.

In your design project, market research should be used to shape your final product and to ensure that your design decisions are founded on public opinion rather than just your own feelings.

Project plan

A **project plan** outlines the key activities in the production process. It shows their order and duration and, on complex projects, the resources required. Once the plan is written, progress can be checked against the plan to ensure that the project is on track and will finish on time and to budget.

Many projects fail because there was either no plan, the plan had serious errors in it or the plan was not followed. A simple project plan can be written as a flowchart; this will show the order that the tasks should be carried out but little else. A better way to show a production plan is with a **Gantt chart**, a type of horizontal bar chart plotted against time. It shows the order that tasks should be done but can also show their dependency, their duration and the amount of resources required.

Objectives

Learn about the role of market research and technological advances in design development.

Discover the benefits of producing a production plan.

Key terms

Consumer survey: a survey to better understand customers' requirements.

Market: a group of existing or future buyers for specific goods or services.

Product development: the development of a product or service from idea to market.

Project plan: an ordered list of the key activities in a project.

Gantt chart: a visual representation of a project plan.

A *Conducting a survey*

∞ links

To discover more about market research visit:

www.which.co.uk

www.marketresearch.com

www.mrs.org.uk

Project Development Schedule

Project steps:	Qtr 1			Qtr 2			Qtr 3			Qtr 4		
	Jan	Feb	Mar	Apr	May	Jun	Jul	Aug	Sep	Oct	Nov	Dec
Explore market need												
Develop concept for product												
Begin development cycle												
Develop GUI												
User interface test evaluation												
Alpha version release												
Quality assurance testing phase 1												
Fix outstanding problems from alpha												
Beta version release												
Quality assurance testing phase 2												
Fix outstanding problems from beta												
Design box and CD labels												
Begin advance advertising campaign												
FCS preparation												
Final quality assurance testing												
FCS release												
Production and packaging												

Key: Development · QA Testing · Marketing · Box art · Milestones

B *A Gantt chart*

Activities

1　Design a customer survey for a product of your choice.

2　Draw up a production plan for one of your projects.

3　Draw a Gantt chart to show the main elements of building a house.

Study tip

- Ensure that you understand the benefits of market research.
- Be able to explain various methods of undertaking market research.
- Be able to produce a project plan and understand the benefits of keeping to it.
- Make sure that you can create or explain a simple Gantt chart.

Summary

Market research is used to steer future products or services.

There are many ways of conducting a survey.

Surveys are flawed as some people lie or change their mind.

A project plan outlines the key activities in the project.

The key to a successful project is staying closely to the plan.

The production plan can be shown graphically on a Gantt chart.

Over the last few decades there has been a revolution in manufacturing processes: computer-aided design (**CAD**) and computer-aided manufacture (**CAM**) have changed industrial processes beyond recognition. Increasingly we are using these methods and techniques in the school workshop too.

CAD/CAM

These terms describe the application of computers in the design and manufacture of a product. Computer-aided design, or CAD, uses sophisticated drawing packages to create geometric models of a product that were historically done by hand in a drawing office (a very time-consuming procedure). CAD has the advantage of dynamic modelling: you can see the product from any angle, examine cross-sectional areas, look at internal stresses or model gas flow. Many problems can be ironed out way before the product reaches prototype stage. CAD has greatly reduced the time needed for design, and consequently saves money. A recent advance is that CAD software is now used actually to *control* the operation of the machine that makes the product. This is known as computer-aided manufacture, or CAM.

CAM has the benefit that an operator can merely download a file, turn the machine on and within a short period of time produce a finished part. The advantages of CAD/CAM are that it is faster, cheaper, easier, more precise, and greatly reduces the period of time from conception of a product to manufacture.

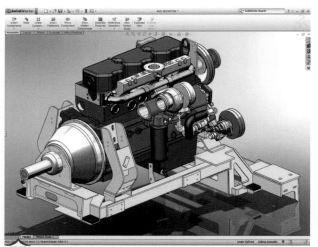

A Solid work software – one of the packages you may come across at school

We can replicate these processes in the school workshop, **Circuit Wizard** to produce PCBs, **Pro/DESKTOP** or **CorelDRAW** driving a laser cutter/engraver, and increasingly nowadays using **SolidWorks** to drive a rapid prototyping machine; these are all identical to commercial manufacturing processes.

Automation

It is difficult to decide where CAD/CAM ends and **automation** begins, as the true definition for automation is any process that reduces the human input in manufacturing a product. Automation is commonly thought of as the use of robots, but applies to any automatic process. However, we must not get confused with the usual meaning of the word 'robot'; **industrial robots** vary greatly, in their size, shape, and function and, in all cases, look nothing like shiny metal men. Many stages in a manufacturing method are very mechanical, and it is these processes we can perform with the machine or robot.

Industrial robots have many advantages:

- They can easily carry large loads.
- They can work 24 hours a day for years on end.
- They can be reprogrammed to do different tasks.
- They are very precise in their movements, replicating tasks over and over with no errors.
- They can operate in any environment, especially those unsuited to human beings.

Industrial robots are ideal for assembly lines as they don't get bored or tired. They are very efficient and improve the quality of the final product.

B   *An industrial robot*

Activities

1   Identify how you use CAD software in your school. List the packages and describe what process it helps you do – for example, Circuit Wizard enables the rapid design, testing and production of printed circuit boards (PCBs).

2   Using the Internet, research the advantages of using CAD/CAM in industry.

3   Do you use CAM in school? Describe the process(es).

Study tip

- Consider what effect you think robots have on society. Try listing the advantages of using robots in manufacturing.
- Can you explain the meaning of CAD/CAM?

links

CAD/CAM in schools exemplar material:
www.cadinschools.org

Virtual tour of graphic package:
www.solidworks.com

History and development of industrial robots:
en.wikipedia.org/wiki/Industrial_robot

Summary

In recent years, computer-aided design (CAD) and computer-aided manufacture (CAM) have revolutionised production processes.

These techniques are increasingly likely to be used in schools as well.

Automation is a development of CAM and includes the use of industrial robots in manufacturing.

Practice questions

Process and manufacture sample questions

1   This question is about the manufacture of a case for a simple car alarm circuit and battery holder.

It is constructed from aluminium with polyethylene end plates.

This drawing below shows one of the rigid polyethylene end plates and the aluminium case (Diagram **A**).

A

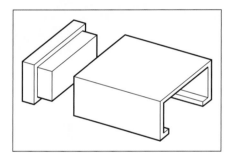

(a)   What is injection moulding? *(2 marks)*

(b)   Why is aluminium a suitable metal for the case? *(1 mark)*

(c)   Sketch an idea that will show how the case idea above could be modified to meet the following specification. The modified case design must include:

- one battery holder
- two LEDs to be used as flashing indicators
- the assembled PCB
- one method of attaching the case to the rear parcel shelf of a car (the shelf is 3mm thick and can be easily drilled)
- methods of mounting the circuit board, battery and LEDs (they must be insulated from the aluminium case).

Assume that two leads leave the case to connect to a hidden switch.
You do not need to include the switch in your solution.

Add notes to your answer to show how you have satisfied the specification.
You may include any other parts or items you feel are appropriate to help complete the case design.

This question is worth 10 marks.

Marks will be awarded as follows:

- quality of communication *(2 marks)*
- quality of notes *(2 marks)*
- method of locating the components *(5 marks)*
- method of fixing the unit to the parcel shelf. *(1 mark)*

2   This question is about pulley systems.

A powered hacksaw needs to move continuously backwards and forwards.

(a)   In Diagram **B** below add a pulley system that will enable the hacksaw operator to choose between two different speeds of operation.

B

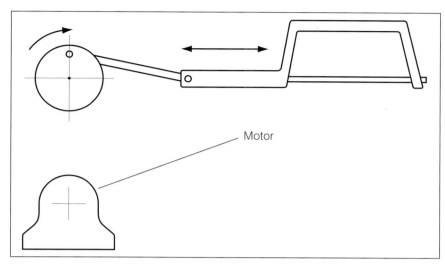

This question is worth 10 marks.

Marks will be awarded for:

- suitability of pulley system   *(4 marks)*
- explanation of how the pulleys are attached to the shafts   *(2 marks)*
- quality of drawing   *(3 marks)*
- notes explaining system.   *(3 marks)*

(b)   As a safety feature it has been decided that the operator needs to press two push-to-make (PTM) switches to start the hacksaw.

Complete the circuit in Diagram **C** by adding a second push-to-make switch and connecting the circuit to the hacksaw motor (M).   *(2 marks)*

C

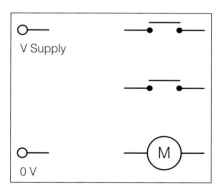

(c)   Explain why the switches should be placed at least 500mm apart on the machine.   *(2 marks)*

GCSE

Introduction

■ What is designing and making practice?

A GCSE grade for Systems and Control Technology is achieved by completing two units:

- Unit 1: A written paper worth 40 per cent of the total marks.
- Unit 2: A coursework project which is worth 60 per cent of the total marks. This project will enable you to show your designing and making skills and therefore is called 'Designing and making practice'. It is assessed to criteria called 'Controlled Assessment criteria' as shown below:

The five criteria together are worth 90 marks.

1 Investigating the design context
2 Development of design proposals
3 Making
4 Testing and evaluation
5 Communication.

Most of your designing and making project has to be done under the supervision of your teacher in the classroom and must meet strict guidelines, hence the name 'Controlled Assessment'. This will mean that there will be restrictions upon what work you are allowed to complete at home.

Remember

- Plan your work carefully.
- Space on pages needs to be used well.
- Your writing or typing size needs to be legible but not too large.
- Keep all research information concise.
- The folder should show your thought process.
- Use system diagrams to analyse other systems and products, and when designing your own.

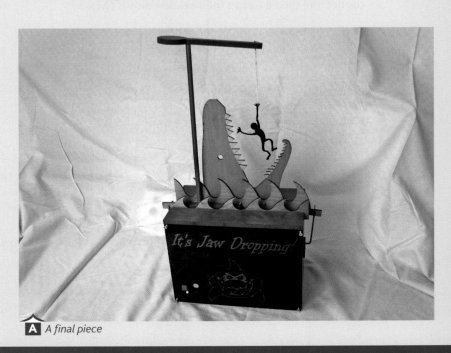

A *A final piece*

A successful project will address all elements of the design process.

What do you need to do?

- With the guidance of your teacher, select a task from a list of design briefs provided by the examination board.
- You must produce a design folder recording the process you have followed, the decisions you have made, reasons for making these decisions and evidence of the 'making you have done'. All the 'making you have done' is also worth a large percentage of the final mark.
- You must produce a final product or system prototype.
- You must produce a final design solution that could be presented to a client or end user for testing.
- You should also model your ideas in a variety of ways.
- All models and products/systems you have made will provide evidence for your final 'making' grade. All your practical work will be assessed by your teacher throughout the project.
- Your folder should consist of 20 sides of A3 paper, or equivalent A4 paper, or ICT equivalent.
- You should spend a maximum of 45 hours on the project.
- Photographic diaries explaining the processes you have used, the problems encountered and tools and equipment used are excellent sources of evidence for your 'making skills'.

Carrying out your designing and making practice

Your folder should follow the design process and be a concise record of what you have done in the process of designing and making your product or system. It should document what you have made and the quality of that making.

In addition to making a complete product or system you need to carry out development work, involving modelling parts or sub-systems of your final product or system.

The functions of the control systems, mechanisms and/or pneumatics you have used need to be understood. You need to comment on how these operate within your product or system.

> **Remember**
> - The quality of making needs to be very good: your product or system should operate in the way you intended in your design.
> - Make models of parts of your final product or system.
> - The functions of the control systems, mechanisms and/or pneumatics you have used need to be understood.
> - Explain how these operate within your product or system.

> **Remember**
> - The quality of your written communication needs to be accurate.
> - There should be evidence of some ICT within your project.
> - Photographic evidence of models and all aspects of the final product should be presented in your folder.

9.1 Research, analysis and criteria

Design brief

Within the specification, design briefs for selection are set in a context to enable focused research. For example:

- **Context:** Automated shop displays and mechanical toys are an increasingly popular way of drawing attention to products and amusing both adults and children.
- **Design task:** You are to design and make an advertising display or automaton. The system you design and make should move or change automatically under certain conditions or at regular intervals.

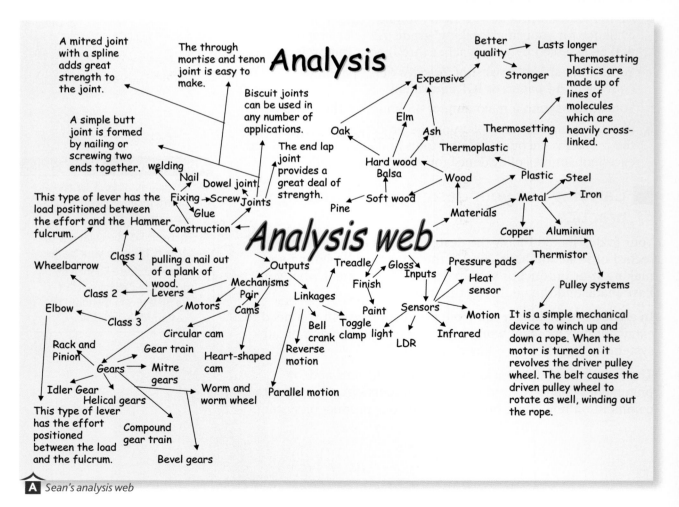

Objectives

Understand how to select relevant research methods to investigate the design brief.

Understand how to carry out detailed research.

Be able to analyse research for writing clear and specific design criteria.

A Sean's analysis web

Research

As you feel the need to investigate further information about the product you are developing, research and its analysis can take place at all stages of the design process.

The following are examples of some of the different types of research that can be carried out and the presentation of the results of the research.

Sean's design brief is to design and make a new advertising display product. He has analysed the brief thinking about the keywords and referring back to the development work he carried out at the beginning of his GCSE course. He has produced this analysis web (shown on the left) for his research work.

Commentary

In Example **A** Sean has undertaken a wide range of research which he has decided to carry out to ensure that he has lots of different information that will ensure that he can write an informed design criteria to meet the brief.

Carrying out practical research is also a good way of including a range of making skills.

Product analysis

A detailed analysis of similar products already available will help you to see a range of features that could be included in the design specification for your new product. Product analysis looks at all relevant aspects of the product including general characteristics (appearance, size, cost, location, user, etc.) and more specific 'system and control' features, such as inputs, process, outputs, mechanism, power supply and safety considerations.

Commentary

Communicating ideas

- This student's work uses arrows well to show linked areas.
- Ideas are communicated effectively, with good use of accurate spelling.

Commentary

In Example **B** overleaf, Will has photographed and shown many existing shop window displays and commented upon them. Note that some of the examples show weak displays; this is useful as it reinforces the stronger shop displays.

Will has also shown some animated displays that are especially relevant to this project, the cobbler and the rotating shelves.

In addition, the photos are relevant and Will has included mature comments around them.

Will used all the information found in the analysis to help him write the design specification for his animated shop window display.

Analysis of research

A good analysis of what the research has helped you find out that you did not know before is essential to help you write your design specification. It is not an analysis of the research methods but the information researched.

Research of existing displays

Not very good use of light; unable to see the window display very well

Rotating glass display shows a lot of products in a small space

Expensive exterior with translucent images on the windows showing the company logo

Back-lit photos of products display clearly and vibrantly

Back-lit well-known brand sign with internet address on

Manikins display the products clearly making it easy for the potential customer to evaluate them.

Big shop sign that is easily seen from a distance

Dark, no lighting

Mechanical man gives the customers an idea of what happens in that shop and it also catches the eye as it's moving

Boring displays of manikins but shop window divided up well with one manikin in each so shop window isn't too busy

It's not very colourful; it's all bland and blends into the back wall

Very very eye-catching shop window as it looks like the window is smashed so people will investigate and see the products inside

Moving shelf attracts people to look at the shelf and the products on it

Very busy, lots of products crowding the window but does catch the eye

B *Analysis of an existing product by Will*

Commentary

Will has commented on his product analysis: this is analysis of research as he is stating what he has discovered and how this information could be relevant to his project.

Design criteria

Design criteria must be written using the new information evaluated from the analysis of research and should also consider the design brief.

Specification

- **Must have an input**
- **Must have an output**
- **Must be inside the shop**
- **Must only take up 50% of the window**
- **Must cost between £100 and £150**
- **Must only run between the opening hours of the store**
- **Must have 2 motion outputs**
- **Must be made from plastic**
- **Must have a modern theme**
- **Sensor must be outside shop**
- **Sensor must be a pressure sensor**
- **Sensor should be mounted on the pavement**
- **Display must run at less than 2000rpm**
- **Must be powered by batteries**
- **Must appeal to the target market**
- **Must take industrial practice into consideration**

 Product specification by Will

Commentary Will has listed the general design criteria and ensured that they will be easy to evaluate against. An easy test is if they can have a yes/no answer. He may have a problem with the following statements as they are subjective and not easily answered:

- must have a modern theme
- must appeal to the target market.

The criteria in the specification do not specify a solution, just a set of features that the solution must have. This ensures that a range of possible solutions can be designed.

Summary

You should be able to choose an appropriate research method for investigating the design brief.

You should understand how to analyse research for writing clear and specific design criteria.

10 Development and investigations to produce a design solution

10.1 Developing design ideas

Design ideas

Initial design ideas should show solutions or part solutions that will satisfy the design specification. It is more important that all of the ideas are noted and conveyed than they are formatted accurately – some of the best ideas started off on the back of an envelope!

Annotated pencil sketches are the best way to record initial design ideas, as these are quick to produce and can be easily edited with an eraser. It is important that, even at the initial design idea stage, you sketch actual materials and components. There is little point in designing a product that you do not understand or cannot explain.

Will and Richard's brief was to produce an animated window display product; they have both chosen to design a solution for a sports shop.

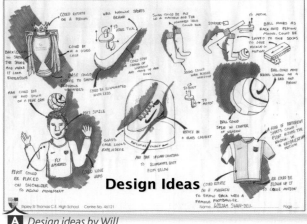

A Design ideas by Will

Commentary

Communicating design ideas

- Good examples of how to set out a design idea page.
- Even though these pages are handwritten and hand-drawn, the presentation is very clear and easily understood.

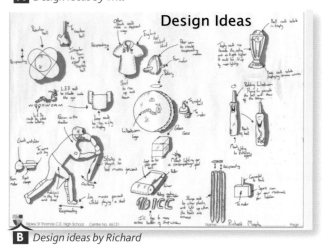

B Design ideas by Richard

AC5 communication

- These are good examples of a sheet of design ideas.
- The sketches are clear and well-annotated.
- Suitable materials, power supplies and mechanisms where required are shown.
- There are also comments regarding cost and construction processes.
- Your folio should show many sketches, even the weak ones!

Developing products

Once sufficient design ideas are sketched and annotated, a selection of the best ideas should be developed further. This involves adding more detail to the original ideas and includes dimensions, materials, manufacturing processes, circuit diagrams, power supplies, components, surface finishes etc.

links

This might be a good moment for some revision on materials and their properties. See pages 10–11.

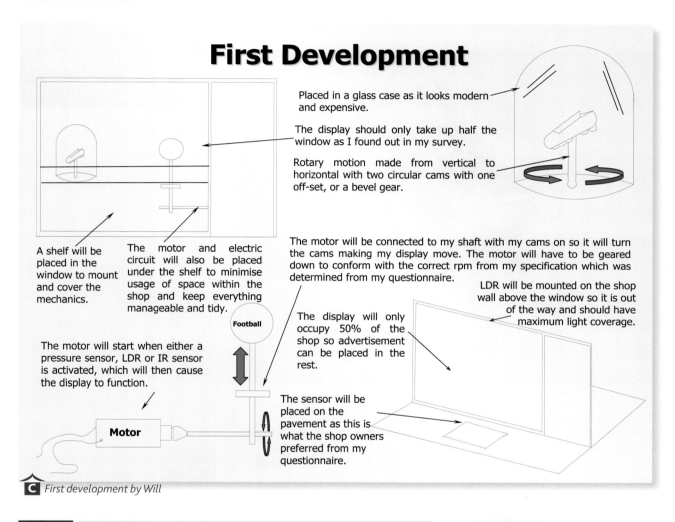

First Development

Placed in a glass case as it looks modern and expensive.

The display should only take up half the window as I found out in my survey.

Rotary motion made from vertical to horizontal with two circular cams with one off-set, or a bevel gear.

A shelf will be placed in the window to mount and cover the mechanics.

The motor and electric circuit will also be placed under the shelf to minimise usage of space within the shop and keep everything manageable and tidy.

The motor will be connected to my shaft with my cams on so it will turn the cams making my display move. The motor will have to be geared down to conform with the correct rpm from my specification which was determined from my questionnaire.

LDR will be mounted on the shop wall above the window so it is out of the way and should have maximum light coverage.

Football

The display will only occupy 50% of the shop so advertisement can be placed in the rest.

The motor will start when either a pressure sensor, LDR or IR sensor is activated, which will then cause the display to function.

Motor

The sensor will be placed on the pavement as this is what the shop owners preferred from my questionnaire.

C *First development by Will*

Commentary
- This is a good example of a development sheet.
- It shows how the product ideas would be constructed.
- It also shows power supplies (motor) and mechanisms (bevel gears).

AC5 communication

Communicating the development of design ideas

- ☞ This is an example of a high-quality presentation.
- ☞ The language is appropriate, easily understood and technical language is used correctly.

Summary

After reading these pages you should be inspired to develop your own design ideas and to communicate them effectively.

Modelling can be used successfully to develop products. Commercial companies produce models, often actual size, to develop their products, e.g. car prototypes, buildings, packaging. The photo below shows a simple robot built from Lego. This is a quick way to check that the system and program will work.

A *Lego robot model*

There are numerous construction kits that can be used to develop and test design ideas. Photo **B** shows a fully-functioning pneumatic model of a production line. It includes a magazine that loads parts onto a rotary table; this moves parts to a processing station and then rotates parts for despatch to a conveyor belt.

Commentary

- Modelling is an excellent way to prove that a design will work.
- Models can be used to check that the final users will approve the design.

∞ links

To find out more about modelling control systems search these suppliers' websites:

Fischertechnik:
www.fischertechnik.de/en

Lego: **www.lego.com**

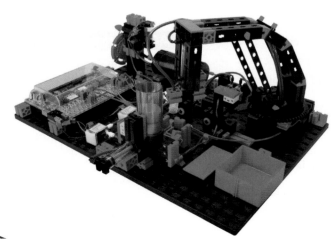

B *Pneumatic model of a production line*

In this example the student has chosen one of their initial design ideas and produced a card model to show how it would work. This is especially relevant in the development of mechanical solutions as it allows experimentation with sizes and dimensions to find the correct ratio and stroke.

DEVELOPMENT

This was the basic design for my shop window that I made out of card first, so then I could develop my design. Here I have chosen to make the design more geometric with more curves as this gives a more contemporary feel to the shop.

I have chosen to fix my display by using a triangular solid piece, this will provide strength but also looks aesthetically pleasing.

These will connect to the cricketers arms and legs to allow them to move when the motor is activated.

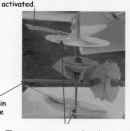

These are the bevel gears that I designed to change the direction of the rotary motion by 90°, to allow me to attach my crank and slider. They are also relatively quiet and are efficient at the job that they do.

This is the main shaft from the motor

These are my two crank and sliders, one is to move the arms whilst the other is for the leg. They are free moving and will not slip, therefore they are a good mechanism to use.

This is the shaft that would be connected to the motor, however the motor would be situated out of sight to prevent the display from looking ugly. Also this shaft would need to be hidden by the door as it looks unsightly.

This shows how big the bevel gears are in terms of the shop display. They are large and take up a lot of room, ergo it may be easier to turn the motor around instead of using bevel gears.

This is my basic cricketer produced from card, however due to it's rigidity, the cricketer wouldn't move if it were to be connected to the crank and slider.

These are pivots that will allow the cricketer to move freely when the sensor has been activated. Also there will be a pivot in the foot of the back leg to prevent the design from moving.

The company logo has been displayed above the shop to let customers what brand they are buying and to advertise the brand.

This is the final card production which shows how the cricketer will fill the shop display. In the background are the mechanisms that will be connected to the man that will make him play a shot when the sensor is activated, consequently activating the motor.

This is my final card model of the shop, it has one large window which allows more room for my display to be placed within. Therefore I will be able to capitalise upon the space by creating an effective display.

C *Development of design ideas by Richard*

AC5 communication

Communication through modelling

Simple modelling is an excellent way to show other making skills.

This student shows excellent use of photographic evidence and ICT.

The technical language used is accurate and the spelling, punctuation and grammar of a high standard.

Commentary

- This is an excellent example of a modelling development sheet.
- The comments on the sheet discuss and explain the model.
- The scale is also relevant as it shows the size of the final product and how the components fit together.
- Students' construction kit models can be submitted as part of their GCSE Controlled Assessment.

Summary

After reading these pages you should be able to choose appropriate materials for modelling your design proposal.

Further development

Once the design idea has been chosen and the model shows that it will work, further development is required to finalise the design. Richard's sheet shows that he is now defining each component that will be required in his final design.

Objectives

Produce development work through experimentation in order to produce a final design solution.

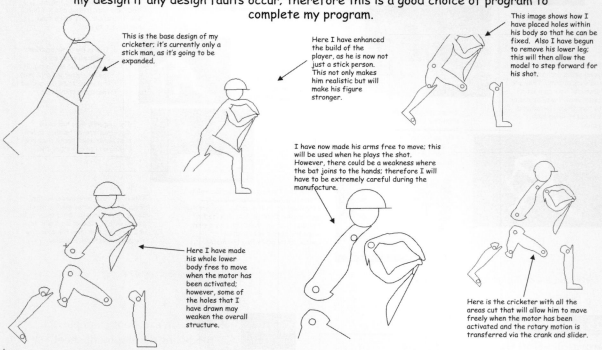

Further Development

I will now begin to design my cricketer using a program called 2D Design, which will allow me successfully to draw the cricketer and then allow me to print it using a laser cutter. Also, as I proceed throughout the designing of my project, I will be easily able to edit my design if any design faults occur; therefore this is a good choice of program to complete my program.

This is the base design of my cricketer; it's currently only a stick man, as it's going to be expanded.

Here I have enhanced the build of the player, as he is now not just a stick person. This not only makes him realistic but will make his figure stronger.

This image shows how I have placed holes within his body so that he can be fixed. Also I have begun to remove his lower leg; this will then allow the model to step forward for his shot.

I have now made his arms free to move; this will be used when he plays the shot. However, there could be a weakness where the bat joins to the hands; therefore I will have to be extremely careful during the manufacture.

Here I have made his whole lower body free to move when the motor has been activated; however, some of the holes that I have drawn may weaken the overall structure.

Here is the cricketer with all the areas cut that will allow him to move freely when the motor has been activated and the rotary motion is transferred via the crank and slider.

 Further development by Richard

Commentary

- This is a good example of a further development sheet.
- It shows details of individual components.
- The comments on the sheet show that the student has a good understanding of the solution.

AC5 communication

Communicating the development of ideas

This student's work demonstrates excellent use of ICT – both CAD and DTP.

The technical language used is accurate and the spelling, punctuation and grammar of a high standard.

Final design

Once the development is complete a final design should be produced. This shows the solution that has been decided upon and can be used to build the final product.

∞ **links**

Pages 106–110 gives you more information about the use of ICT and CAD in schools.

Final Design

This is my final design that I have designed used CAD; this shows how I will fix my design together within my shop window. It also shows how my model will look in comparison to my shop window.

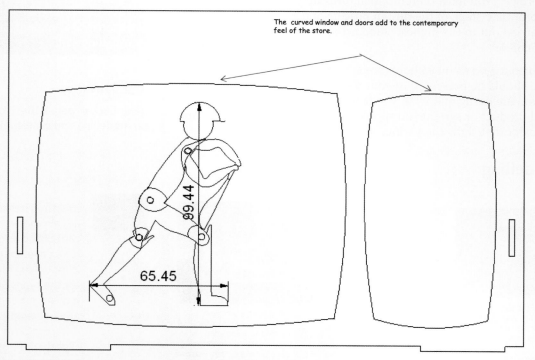

The curved window and doors add to the contemporary feel of the store.

99.44

65.45

B *Final design by Richard*

Commentary

- This is a good example of a final design sheet.
- It shows details of the arrangement of parts in the final product.
- It shows overall dimensions of the main parts.
- There is no requirement to draw every part out as there is insufficient time; it is better to draw a few parts well than to draw all parts poorly.

Summary

After reading these pages you should feel confident about finalising your design proposal.

11.1 Making and modelling

The making that you carry out accounts for a large proportion of the marks available for your Controlled Assessment GCSE Systems and Control project. You must therefore demonstrate your making skills at every appropriate stage of the design process and ensure that they are always carried out to the highest standard of accuracy and quality that you can achieve.

You are encouraged to take photographs of making at every opportunity and these should be presented in your design folder with explanations of what you are doing and what you have found out. When making your final product or system you should also try to include photographs in your folder that show the various stages of making.

Objectives

Understand how to produce a final outcome that demonstrates making and modelling skills.

⊗links

Refer back to pages 16–17 to read up on modelling and prototyping.

Modelling

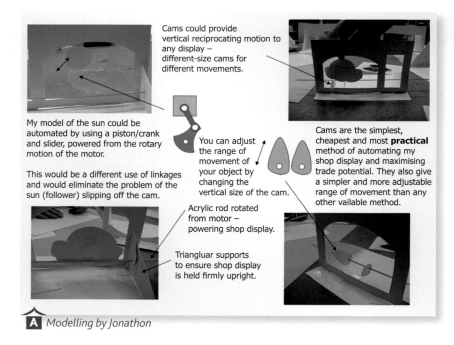

Cams could provide vertical reciprocating motion to any display – different-size cams for different movements.

My model of the sun could be automated by using a piston/crank and slider, powered from the rotary motion of the motor.

This would be a different use of linkages and would eliminate the problem of the sun (follower) slipping off the cam.

You can adjust the range of movement of your object by changing the vertical size of the cam.

Cams are the simplest, cheapest and most **practical** method of automating my shop display and maximising trade potential. They also give a simpler and more adjustable range of movement than any other vailable method.

Acrylic rod rotated from motor – powering shop display.

Triangluar supports to ensure shop display is held firmly upright.

A *Modelling by Jonathon*

Commentary
Jonathon has made a scale model in card of the display he is developing. He is modelling ideas for the types of movement he requires and has tried two different cams of slightly different size to work out which one will give him the movement he desires. His model, although quite basic, is well made with carefully cut-out shapes and serves its purpose well.

Jonathon is modelling ideas for a shop advertising display; he is developing ideas for the types of movement he requires.

Using photographs

Systems & Control

Project Diary

Photo

Problems Encountered/Changes Made

Levers to move the donkey's legs.

My initial plan for this automaton was to use the solenoid direct to the donkey's legs but the main problem is that solenoids only have around 10-20mm of movement.

This lever will extend the length of the movement for the other side of the pivot. It is a rather simple lever but it is the best design that will give the most effective outcome.

I may need to add a small spring to the lever so it doesn't jam. It will not need to have much tension as it will just be there as a backup in case the weight of the donkey's legs don't knock the solenoid back into place.

B *An example of photos used in a project diary by Phil*

Phil is making an automaton controlled by an electronic circuit.

Summary

Reading these pages will help you understand how modelling can be used for both designing and testing design ideas.

Remember

Photographs are a good way of recording the making process and can be used to help explain decisions made during making.

Commentary

In his project diary, Phil outlines the stages of making and explains how he overcame any problems he encountered. On the page shown he explains why a lever was added to the solenoid he used as an output. The page helps demonstrate that he has worked independently and shows that he has set himself challengingly high standards for the final outcome.

Quality control

In order for your product or system to work as intended it is very important in Systems and Control Technology to make sure that the quality of what you are making is at the highest standard you can possibly achieve. To do this you should build into your production plan the quality checks you are going to make and when you are going to make them.

Objectives

Understand the need for quality control checks throughout the project.

Understand that outcomes need to be suitable for a target market.

Commentary | In this example of a production plan, Harry has used a flow diagram to show the sequence of operations involved in making his project. This section shows the stages in making a printed circuit board and the diamond-shaped decision boxes describe quality checks that he will make at various stages of production to make sure the quality of the final board is of a standard for the circuit to work correctly.

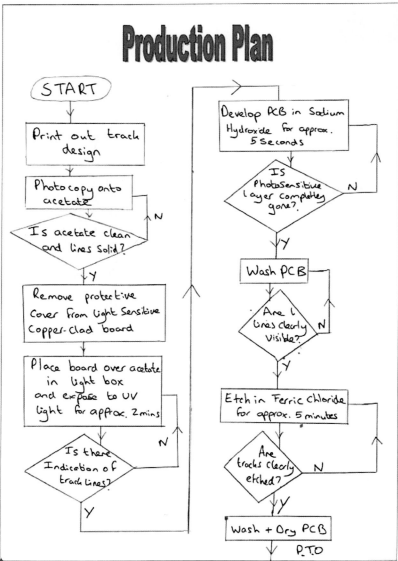

Production Plan

START

Print out track design

Photocopy onto acetate

Is acetate clean and lines solid? N

Remove protective cover from light sensitive copper-clad board

Place board over acetate in light box and expose to UV light for approx. 2 mins

Is there indication of track lines? N

Develop PCB in Sodium Hydroxide for approx. 5 seconds

Is photosensitive layer completely gone? N

Wash PCB

Are lines clearly visible? N

Etch in Ferric Chloride for approx. 5 minutes

Are tracks clearly etched? N

Wash + Dry PCB

P.T.O

A *An example of a production plan with quality checks highlighted by Harry*

∞ links

Refer back to pages 37–39 to remind yourself how to draw systems and control flowcharts.

Quality counts

If your product or system is going to meet the needs of the target market, then the final quality and operation of the product is really important. To make sure that your system works then any control circuits and systems need to be soldered and in working order, any mechanical or pneumatic systems need to be well made and mounted accurately and securely. This can be achieved in many different ways; you may want to mark out and cut accurately any parts with the workshop tools available to you, or you may have access to and prefer to use computer-aided design and computer-aided manufacturing systems. Whatever approach you take, make sure everything is carried out to the highest degree of accuracy you can manage.

The drawings Phil photographed for his project folder shown in the previous spread were for the automaton below. He has made good use of laser cutting to ensure the accuracy needed for putting together this fully-working example.

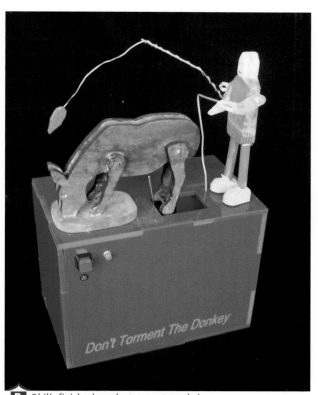

C *Phil's finished product: an internal view*

B *Phil's finished product: an external view*

Commentary	In this example Phil shows some high-level making skills and has evidence in his portfolio of the original CAD work used to make the product. The overall presentation of his work is very good with wires, circuits and batteries carefully routed and mounted to help make sure the product functions as intended.

AC5 communication

Include good, annotated evidence of CAD work.

Examples of making

These pages display a variety of finished products showing accuracy in making, good quality control and high-quality finish.

D *David's finished product: an external view*

H *Andrew's finished product: an external view*

E *David's finished product: an internal view*

I *Andrew's finished product: an internal view*

Commentary

These products and systems were made as the final outcome after the completion of the design process and show a range of systems including a golf putting return system, a table tennis ball launching system and an automaton. They all show the degree of accuracy that comes from having a quality control system that ensures everything fits together and works as the students intended.

Excellent photographic evidence showing high-level making skills and finish.

Summary

After reading these pages you should understand the importance of building in rigorous quality checks when production-planning.

12 Testing and evaluation

12.1 Testing and evaluating

Evaluating and testing design ideas

The controlled assessment must describe the decisions you make in developing your system. It is therefore important that evaluating and testing is carried out throughout the design process and should not be viewed as something that only happens at the end of your project. After coming up with design ideas you will need to refer back to your specification and comment on how well your ideas match your intentions. If you are designing an electronic system you will need to model and test your ideas. If you are designing a mechanical or pneumatic system you will need to test and evaluate your ideas.

Kasey is designing the electronic system to open and close curtains or blinds in the home. In her research she discovered rotary motion could be turned into linear motion using a device called a linear actuator, and in her specification she outlined the need for the current to the linear actuator to reverse in order to get the blinds to open and close. On this page she shows her electronic system design as a block diagram and models her idea using computer-modelling software.

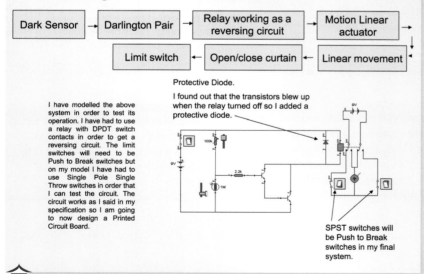

A A design and computer model of an electronic system by Kasey

Objectives

Understand the importance of testing and evaluating throughout the designing and making process.

links

Refer back to pages 80–81 to remind yourself about evaluation techniques and quality control through testing.

Remember

If you have produced a diary to accompany the making process, in which you have talked about any problems you may have encountered and how you overcame them, this will also be good evidence for criterion 4.

Commentary Kasey has considered what she said in her specification regarding the function of the electronic system and produced a model to test out her system idea. The model shows that her electronic system design has the potential to operate successfully and she has made some simple comments about the types of switches used in the system.

AC5 communication

The decisions are communicated in a clear and coherent manner using some technical language. Excellent communication is achieved through very clear and concise diagrams.

Jonathon is designing the mechanism and considering the types of movement he requires in his advertising display. He is drawing from some of the information he found out in earlier research and he is explaining some of the decisions he is making about his design.

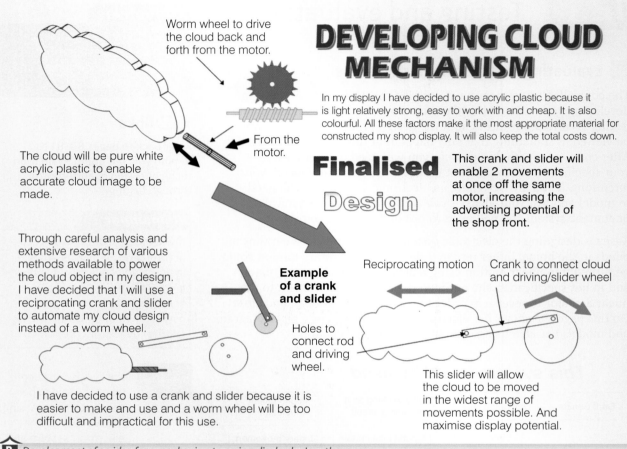

Worm wheel to drive the cloud back and forth from the motor.

DEVELOPING CLOUD MECHANISM

In my display I have decided to use acrylic plastic because it is light relatively strong, easy to work with and cheap. It is also colourful. All these factors make it the most appropriate material for constructed my shop display. It will also keep the total costs down.

From the motor.

Finalised Design

The cloud will be pure white acrylic plastic to enable accurate cloud image to be made.

This crank and slider will enable 2 movements at once off the same motor, increasing the advertising potential of the shop front.

Through careful analysis and extensive research of various methods available to power the cloud object in my design. I have decided that I will use a reciprocating crank and slider to automate my cloud design instead of a worm wheel.

Example of a crank and slider

Reciprocating motion

Crank to connect cloud and driving/slider wheel

Holes to connect rod and driving wheel.

This slider will allow the cloud to be moved in the widest range of movements possible. And maximise display potential.

I have decided to use a crank and slider because it is easier to make and use and a worm wheel will be too difficult and impractical for this use.

B *Development of an idea for a mechanism to go in a display by Jonathon*

Commentary

Jonathon has undertaken appropriate evaluation of his designs on this page and has gone some way to explaining his decisions. Full justification of the chosen idea and possible areas for development are considered. This is further expanded and modelled later in the project. Jonathon has mentioned links to his research pages demonstrating a good understanding of the design process.

Remember

Always explain the choices you make with your designs and refer back to your design criteria as often as possible.

Summary

You should be able to test and evaluate as appropriate throughout the designing and making process.

12.2 / Evaluating outcomes

Testing the final solution

You will need to test your final product or system to find out how successful your design has been. To do this you will need to try it out yourself, get other *end users* or clients to try it out. You should make notes of people, thoughts and observations, but in order to provide evidence of the testing of your design you could produce a list of closed questions to ask yourself and others after the design has been tested. The results of these questions could be collated into an easily readable form by using pie charts or bar charts to present the results. These results can then be analysed in the final evaluation and help to make recommendations for future modifications.

Objectives

Understand the importance of testing all aspects of the final outcome against the design criteria and specification.

Be able to produce an outcome that is suitable for a target market.

Remember

Always try to leave plenty of time to carry out this section.

Remember

In systems and control technology, it is important that your photographs do not just show the outside of your product or system but also show any electronic circuits you have made and any mechanical or pneumatic control systems that may be hidden inside.

Testing Questionnaire

After making my *Table Tennis Ball Launcher* I tested it myself by trying it out and I got friends and members of the school table tennis club to try it out. To help me evaluate the success of my project I asked them the following questions. I was handed back 20 completed questionnaires including my own.

		YES	NO
1.	Did the *Table Tennis Ball Launcher* turn on every time? ...	☐	☐
2.	Did it launch balls every time?	☐	☐
3.	Did it get the balls over the net?	☐	☐
4.	Did it just fire one ball at a time?	☐	☐
5.	Did all 3 launch settings work?	☐	☐
6.	Would you use it again to practise?..............................	☐	☐
7.	Could you carry it easily?.......................................	☐	☐
8.	Did you think it launched enough balls?..........................	☐	☐

I asked 20 people and here are some charts to show my results.

Commentary

Kelly has devised a list of questions that will lead the end users to carry out appropriate testing in order to evaluate her table tennis ball launching system. She has carefully chosen a range of people to carry out the tests in order to ensure a range of opinion from a variety of end users.

A *An example of a testing questionnaire by Kelly*

Kelly has made a table tennis ball launcher and has asked a number of people to test out. Other than herself she has asked some friends to try it and has also asked some members of her school table tennis club to try it. She has thought of a number of *closed* questions to ask the people who tested her product and given them to them in the form of a questionnaire. Kelly has then presented the results to the questions in the form of a series of bar charts.

The final evaluation

The final evaluation should discuss how successful your final design solution has been. You will need to comment upon a variety of issues and include the following:

- a photograph or photographs of the final solution
- comments regarding how well the final solution matches your design specification
- what potential end users or experts said
- whether further modifications need to be made and if so what they would be.

Evaluation

Specification points	Evaluation comments	Modifications
• The shop display must use the colour blue.	• My shop display is made out of two shades of blue plastic.	• I had to use some of my spare plastic to make a back piece for the mechanisms to be attached to. This was much easier than sticking them down to the base of the shop display and so this has made them more secure. It also means I have free-flowing movement with things such as my crank and slider mechanism, which is vital for the smoothness of my mechanisms.
• The shop display must be suitable for a range of ages.	• I would say that my shop display would appeal to a large range of ages.	
• The display must be between 50% and 75% automated.	• The display is about 50% automated, which fits my specification.	
• The display must cost less than £100.	• The display cost £15 overall to make, which is well under £100.	
• The display must run for between 30 and 60 seconds at a time.	• My display runs for 4 seconds at a time.	• After seeing my display working I have realised that 30 seconds is an extremely long time and so I have shortened the time to just 4 seconds. This should be a much more realistic time and should also stop the battery from running out too fast as it would if the mechanisms ran for 30 seconds at a time.
• The display must be powered by a 9v battery.	• The display is powered by a 4.5v battery.	
• The display must be made out of plastic.	• The display is made out of plastic as it is supposed to.	
• The display must incorporate some sort of pressure pad.	• The input is a 'push to make' switch which is basically a pressure pad.	• A 9v battery would have been too powerful for our display and would have made the mechanisms run too fast and so I am now using a 4.5v battery which allows the mechanism to run at a slower speed. This should allow people to get a better look at the mechanisms and the rest of the shop display.
• The output must be movement of some kind, meaning a mechanism must be involved.	• The output is two moving mechanisms so there is movement.	
• The display must run between 2000 and 5000 revs per minute.	• The display runs at 2000rpm.	
• The shop display must have a fairly modern design.	• The design is fairly modern.	
• The display must be able to run when the shop owner sees fit such as during busy parts of the day.	• The display has another switch to allow it only to run when switched on by the shopkeeper.	• My display only has one mechanism because it was difficult for me to have both mechanisms running at the same time. I tried using a pivot to join the two mechanisms but no matter what I did the pivot made the mechanisms only work for part of the cycle. The use of two mechanisms caused problems in that the motion which should have been smooth wasn't. So I decided to settle for one mechanism which is now very smooth.
• The display must have an aesthetically pleasing design.	• I believe that the design looks quite good.	
• The shop display should incorporate two mechanisms.	• My design incorporates one mechanism.	
• The shop display must have supports to keep it upright.	• The display has two supports which hold it upright.	
• The shop display and mechanisms must be fixed in place to stop them from being damaged.	• My display and mechanisms are stuck in place using either glue or stoppers and plastic rods.	

B *An example of a final evaluation part 1 by Lewis*

Lewis has completed all aspects of making and testing his final shop display and is presenting here the first part of his final evaluation. In this section he is showing the points from his design specification and commenting upon how successful he has been in achieving them and also listing the modifications he had to make for the successful completion of the project.

EVALUATION

3rd PERSON PRODUCT EVALUATION

The display is reasonably simple but very effective. The shop display is very eye-catching in its own right. The display is eye-catching but I believe that it could be more attractive to the eye if more varied colours were used.

The way in which the display is made has been thought out and that is noticeable but I think the sun is going to be unreliable due to the set-up. The sun will slip off the pear cam although the design is very good. The sun rising is simple but effective. The cloud is being animated by using a crank and slider. This is a simple mechanism but has been used very cleverly. The shop name is used effectively due to the colours which are bright and cheerful colours which will entice the customer to enter the shop.

Only using one motor could have impeded his design but he has overcome this by using gear trains which will connect the two outputs together. The idea of having two outputs is very clever and linking them up with the singular motor has been thought out cleverly.

By Gavin

FINAL COMMENTS ON DESIGN PROJECT

I manufactured my display again I would change the following things:

- I would ensure all measurements were accurate.
- I would make sure that all the mechanisms worked together properly.
- I would get the right size gears to drive my display and ensure they were the right size.
- I would make sure that the gears meshed properly and they were the right size and number of teeth to reduce th motor to the desired RPM.
- I would use the plastic cement for gluing everything in my display - because the hot glue did not stick effectively and tended to break off.
- I would have a specific place for my circuit board to go (hidden) instead of having it hanging below my display.

Commentary

Gavin has finished off his evaluation by giving an account of a third-party opinion of his design. After testing his display and evaluating its success on the previous page of his project, he gives detail of modifications that would be necessary if the product were to be re-produced.

The photograph of the final product only shows one view, and it would be useful to see photographs of the circuit and how it is mounted, of the gear system Jonathon describes and of the crank mechanism mentioned in the third-party evaluation. It is important to illustrate all the aspects of the project described in the evaluation.

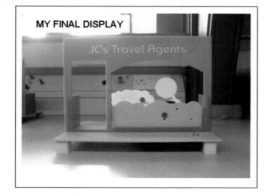

MY FINAL DISPLAY

JC's Travel Agents

 An example of a final evaluation part 2 by Gavin

Gavin has presented the final part of his evaluation here. In this section he has presented the views of a third person and sums up by saying what further modifications he would make if he were to make his display again.

Summary

You should now understand why it is important to test all aspects of the final outcome against the design criteria and specification, and be able to produce an outcome that is suitable for a specific target market.

How long should the project be?

The design folder should consist of approximately 20 sides of A3 paper or A4 equivalent or ICT equivalent. The space on each page should be used well and writing should be concise and accurate.

How much time should be spent on the project?

The complete project, including all design work and all making work, should take 45 hours.

Can I complete any of the work at home?

Your project will be closely supervised by your teacher, with most of the work being carried out in school. This will mean that there will be restrictions upon what work you are allowed to complete at home and your homework tasks are likely to be related more to planning what you will need to be effective in the next D&T lesson or related to the written paper. It will, of course, mean that for every lesson you will need to ensure that you are making full use of the time available.

Can I write my own design brief?

No. You must use a brief which has been set by the Awarding Body (AQA) and given to you by your teacher. You may be given several briefs to choose from.

What does 'target market' mean?

The target market is the specific or particular group of people at which your product or system is aimed. The target market might be an individual in a one-off system that helps someone in a very specific way, or a group of people where the product or system has to meet the needs of everyone in the particular group.

How much research do I need to produce?

Research can occur at any point in the design process and should always be relevant. A lot of research is done at the beginning of a project, often unnecessarily. A maximum of two sides of A3 at the beginning of your project is sufficient, supplemented by other research at relevant points.

Do I need to sketch my design ideas?

After a design specification has been finalised design ideas should be generated. This can be in the form of sketches or computer-generated images. When producing design ideas it is important to thoroughly annotate the ideas to show the types of movement you hope to achieve, dimensions, possible materials etc. Sketching is good preparation for the examination but not a mandatory requirement for the Controlled Assessment.

How many design ideas should I make?

There is no exact number required but at least three would be a recommendation. Remember initial ideas should be presented as system block diagrams describing possibilities for inputs, processes and outputs.

How many products or systems should I take to the developmental stage?

One product or system should be taken forward to the developmental stage after carrying out analysis and evaluation of your generated ideas. It is during this stage that you may want to model your ideas for the sub-systems in your product or system.

Does my final product or system have to work?

It is expected that your final product or system does work. However, it is understood that the initial prototype may not work completely as intended; if this is the case then you should explain this in your final evaluation.

You need to be able to work independently and be organised throughout making activities. The final product or system you make must be accurately made with the potential to work as intended and be a quality item. Throughout making you need to show a thorough understanding of the components that you are using. If you are using a lot of kit components then your system should be more complex with perhaps a greater number of sub-systems being used.

Do I need to include photographic evidence?

Photographic evidence of the finished outcome must be produced. It is also strongly recommended that photographic evidence should be included at various stages of making. It is very good practice to show evidence of the completed design ideas and the different making stages of the final product – perhaps in the form of a production diary.

How do I present of the folder?

It is important to present your work to the highest standard possible. When marking your work your teacher will be looking for good spelling, punctuation and grammar. It is important to present your work in a clear and coherent manner. Make sure that you use correct technical language when explaining the components and systems you are using.

Does my project need to be word processed?

No, however criterion 5 assesses your written communication skills and word processing could ensure your work has text that is legible, easily understood and shows good spelling, punctuation and grammar.

Who will mark my work?

Your teacher will assess everything you have done which includes all work in your design folder and all the 'making' you have done. A moderator from the Awarding Body (AQA) will request some or all of the Systems and Control Technology work from your school and will re-mark it. You will then be awarded a mark for your Designing and Making Practice which will be added to the mark you get for the written examination. Your grade will then be calculated.

Glossary

A

Aesthetic: visual appearance.

Alloys: metals that are mixed with other elements to change the properties of the metal.

Analogue: any value within a range.

Armature: the moving iron part of a relay.

Automation: mechanical or electronic equipment that minimises human labour in the workplace.

B

Base, collector, emitter: the three legs of an NPN transistor.

BASIC: a simple programming language used in schools and colleges.

Batch production: small quantities of a similar product made at the same time.

Bearing: a device that reduces the force of friction, thereby increasing efficiency and reducing wear.

C

CAD: computer-aided design.

Calibrate: to check, adjust or determine by comparison with a standard.

CAM: computer-aided manufacture.

Chain and sprocket: a pair of toothed wheels connected by a chain.

Closed questions: questions which give a limited choice of answer.

Compound gear train: several pairs of interlocking gears with two gears mounted on the same axle.

Conductor: a substance that conducts electricity, e.g. copper.

Consumer survey: a survey to better understand customers' requirements.

Continuous flow production: the product is made continuously over a period of time, which might be years.

Copyright: legal protection for written or recorded work.

Current: the flow of electric charge.

D

Design: a unique style of existing products.

Digital – either ON or OFF, 1 or 0 state: the condition of an input or output.

Discrete component: a single electronic component.

Drive shafts: a tube or a rod used for transmitting power.

Dry joints: soldered joints that do not provide an electrical connection.

Dual in line (DIL): a type of chip consisting of two lines of legs 0.1 inches apart.

E

Effort: the input of a machine.

Electrical potential: another term for voltage.

Electromotive Force (EMF): the energy measured in volts, that is supplied by a source of electric power in driving a unit charge around the circuit.

Environmental: belonging to the natural world: land, sea, air, plants and animals.

Etch: to cut into or remove metals with strong acid.

F

Feedback: the sending of information back into the system, e.g. to say if a limit has been reached.

Field effect transistor (FET): a type of unipolar transistor.

Fixture: a work holding or positioning device that is fixed to a machine bed.

Flowchart: a chart that uses standard symbols to describe a sequence of operations.

Forming: changing the shape of materials.

Frequency: the number of repetitions per unit time of a complete waveform.

Friction: a force that resists the motion between two surfaces.

Fulcrum: a fixed or movable pivot.

G

Gantt chart: a visual representation of a project plan.

Gate, source, drain: the three legs of an FET.

Gear: a toothed wheel of varying diameter.

Gear ratio: the relationship between the driver gear and the driven gear.

H

Heat sink: a device used to absorb heat away from a component to prevent damage.

Hertz: the SI unit of frequency. One hertz is one cycle per second.

I

Idler gear: a gear that reverses the direction of the second gear without changing the gear ratio.

Industrial robot: a re-programmable manipulator controlled by a computer.

Input: information put into a system for processing.

Insulator: a substance that does not conduct electricity, e.g. rubber.

Interface: a device that connects two independent circuits or systems.

Interference fit: a method of mounting gears, pulleys and sprockets onto a drive shaft.

Integrated circuit (ICs): a miniature electronic circuit, on semiconductor material.

J

Jig: a work-holding or positioning device that is not fixed to a machine bed.

L

Latch: a circuit that stays ON even when the original trigger signal is removed.

Lever: a simple machine used to transmit force.

Licensing: a legal framework that allows inventors and manufacturers to agree terms when working together.

Light-emitting diode (LED): a semiconductor that gives out light.

Linear motion: movement in a straight line.

Linkages: a series of linked and pivoted bars that produce various forms of motion.

Liquid crystal display (LCD): a low-powered flat panel display.

Load: the output of a machine.

Logic gate: performs a logical operation on logic inputs and produces a single logic output.

Logic: rules that determine a process; known as Boolean logic.

Low voltage lamp/bulb: an electric light in which a filament is heated.

Lubricant: a substance placed between two surfaces that reduces friction.

Lubricate: to make slippery.

M

Maintenance: the work of keeping a product in good condition.

Market: a group of existing or future buyers for specific goods or services.

Mark/space waveform: the ratio between the high and low part of a waveform.

Mass production: a production process in which large numbers of the product are made in stages and finally completed as a whole.

Mechanical advantage (MA): the ratio of the output force when compared to the effort force.

Mechanism: a device that can be used to convert one type of motion into another.

Microcontroller: a type of microprocessor, a computer on a chip.

Modification: making changes to improve the system.

N

N: abbreviation for newtons, the units used to measure force.

Non-renewable energy: energy resources that are not replaced immediately.

NPN transistor: a type of bipolar transistor.

O

Obsolescence: the period of time after which a product ceases to function.

Oscillating motion: backwards and forwards swinging movement.

Output: the information produced by a program or process.

P

Patent: official registering that protects new inventions.

PCB: abbreviation for printed circuit board.

PCB mask: a transparent film with image of circuit placed onto it.

Peripheral interface controller (PIC): a microcontroller that can be programmed to react to various input sensors and give a variety of outputs.

Polarised: refers to components with a positive and negative leg that must be placed in a circuit the correct way.

Pollution: the introduction of contaminants into an environment.

Polymorph: a material that can easily be formed or moulded at quite low temperatures.

Populated: components placed and soldered into PCBs.

Potential divider: a resistor connected in series between +ve and 0v supply of a circuit, used to divide the supply voltage.

Printed circuit boards (PCBs): Boards onto which components are soldered with copper tracks connecting them together.

Process: a computing operation.

Product development: the development of a product or service from idea to market.

Production line: mechanical track in a factory upon which products are assembled.

Product life cycle: the various stages in a product's lifespan.

Program: a set of coded instructions that enables a computer to perform a desired sequence of operations.

Project plan: an ordered list of the key activities in a project.

Properties: characteristics of materials, such as hardness and tensile strength.

Prototype: one single item, sometimes built to test a product.

Pulley and belt: a wheel with a grooved rim for a rope or band.

R

Real world view: a computer-generated image from the component side of the PCB, showing the position of components.

Reciprocating motion: movement from side to side, or up and down.

Recycling: using previously-used materials to make fresh products.

Relay: a device that provides a motor or a solenoid with its own power supply.

Renewable energy: energy generated from natural resources which are naturally replaced.

Resistance: the amount that an object limits an electric current.

Resistor: a component that slows down the flow of electrons.

Risk assessment: a systematic way of identifying potential health and safety hazards and minimising potential harm.

Robot: a device that is operated by remote control or computer program.

Rotary motion: movement round and round, like the wheel of a watermill.

rpm: the speed something turns measured in revolutions per minute.

S

Sensor: a transducer that converts one type of energy to another, e.g. light to electricity.

Service: a routine inspection and the carrying out of maintenance activities.

Service interval: the time between services.

Seven segment display: seven LEDs positioned to form a figure of eight.

Short circuit: a type of electrical circuit failure.

Signal: a message between devices.

Simple gear train: two gears meshed together (with or without an idler gear).

Solenoid: a current-carrying coil of wire that acts like a magnet when a current passes through it.

State: the condition of an input or output.

Sub-routine: the operation of a sub-system within a system.

Sub-system: a complete system that forms part of a larger more complex system.

Sustainability: using materials that can be replenished.

Sustainable design: designed in an environmentally friendly way and so that it will last.

Switch: a device to break or connect an electric circuit.

T

Tensile: referring to a stretching force.

Testing: gathering information about the success of the system.

Thermoplastics: plastics that can be reshaped by reheating.

Threshold voltage: the voltage that a transistor switches ON, usually 0.6V.

Toggle clamp: a clamp that uses a locking mechanical linkage called a toggle mechanism.

Torque: a twisting force.

Trademarks: symbols or logos that distinguish products from others.

Transistor: an electronic component that can be used as a switch or amplifier.

Truth table: a table that shows a system's inputs, outputs, and all of the possible states.

V

Viscosity: the thickness of a liquid such as a lubricant.

Voltage: the difference of electrical potential between two points of an electrical circuit.

Index

The authors and publisher are grateful to the following for permission to reproduce the following copyright material:

Chapter strips: Chapter 1, © Phil Degginger/Alamy; Chapter 2, © J Marshall – Tribaleye Images/Alamy; Chapter 3, Chapter4, Chapter 8, Chapter 9 iStockphoto; Chapter 5 © Sandra Baker/Alamy; Chapter 6, Fotolia; Chapter 7 © Troy GB images/Alamy;.Chapter 10, Innova Systems; Chapter 11, Chapter 12 Nathan Allan Photography.

Page 5 Nathan Allan Photography; Page 9A Fischertechnik; 1.2A, 1.2B C.R. Clarke; 1.3B Good Hand Inc; 1.4A Rex Features/Nils Jorgensen; 1.4B Crocodile Technology; 1.5A Alamy/Eddie Gerald; 1.5B www.bcae1.com; 1.5C Techsoft; 1.6A, 1.6B Rapid Electronics Ltd; 1.6C Nathan Allan Photography; 1.7A, 1.7B Rapid Electronics Ltd; 1.7C Crocodile Technology; 2.1A Wikipedia; 2.1B Wikipedia/Iainf; 2.1C Rapid Electronics Ltd; 2.1d Science Photo Library/Andrew Lambert Photography; 2.1E Wikipedia/John Maushammer; 2.2A, 2.2B Rapid Electronics Ltd; 3.1A. 3.3B iStockphoto; 3.1B Rapid Electronics Ltd; 3.2A, 3.2E, 3.2F, 3.2G, 3.2H, 3.2I Rapid Electronics Ltd; 3.4A, 3.4B iStockphoto; 4.2B www.autopetfeeder.co.uk; 4.4I iStockphoto; 4.5B Rapid Electronics Ltd; 4.6A Fotolia; 4.6C iStockphoto; 4.7E Rapid Electronics Ltd; 5.1B Flickr/tompagenet; 5.1C Alamy/Jaileybug; 5.1E, 5.1G iStockphoto; 5.1J Fotolia; 5.2D www.allproducts.com; 5.3B, 5.3D iStockphoto; 5.3F, 5.3I Fotolia; 5.5A Wikipedia/Glenn McKechnie; 5.5B Wikipedia/Gregory David Harington; 5.5C Wikipedia/Arthur Clarke; 5.5D, 5.5E iStockphoto; 5.6A, 5.6C Rapid Electronics Ltd; 5.6B Electrolux; 5.6E Alamy/David J. Green – electrical; 5.7B, 5.7C iStockphoto; Page 76A Getty Images; Page77B Dualit; Page 77C iStockphoto; 6.1B, 6.1C, 6.1E Baxi; 6.1D Grundfos; 6.2C iStockphoto; 7.1B Alamy/The Print Collector; 7.1C, 7.1D iStockphoto; 7.2A Corbis/Gaetano; 7.2B Goodyear; 7.2C Wikipedia; 7.2D Dualit; 7.3A Dyson; 7.3B logos appear courtesy of BT Cellnet, Adidas Ltd, Marks & Spencer; 7.4A, 7.4B, 7.4C, 7.4D iStockphoto; 7.4E US Department of Defense; 7.5B Fotolia; 7.6A, 7.6B, 7.6C, 7.6D iStockphoto; 7.7B Fotolia; 7.7E iStockphoto; page 99A iStockphoto; Page 99B Getty Images; 8.1A Getty Images; 8.1B, 8.1C iStockphoto; 8.3A iStockphoto; 8.4A Innova Systems; 8.4B iStockphoto; 10.2B Fishertechnik; Page 132A, 133B, 134C courtesy of Nathan Allan Photography.

Artwork and photographs supplied for Unit 2 courtesy of students at Burnside Enterprise College, Belvidere College, Ripley St Thomas, Lancaster, Great Barr School, Birmingham.

Additional artwork by Fakenham Photosetting Ltd and Angela Knowles.

Every effort has been made to contact the copyright holders and we apologise if any have been overlooked. Should copyright have been unwittingly infringed in this book, the owners should contact the publishers, who will make corrections at reprint.

The Controlled Assessment tasks in this book are designed to help you prepare for the tasks your teacher will give you. The tasks in this book are not designed to test you formally and you cannot use them as your own Controlled Assessment tasks for AQA. Your teacher will not be able to give you as much help with your tasks for AQA as we have given with the tasks in this book.